Sathyajith Mathew

Wind Energy

Fundamentals, Resource Analysis and Economics

with 137 Figures and 31 Tables

Springer

Dr. Sathyajith Mathew
Assistant Professor
& Wind Energy Consultant
Faculty of Engineering, KCAET
Tavanur
Malapuram, Kerala
India

E-mail : windbook@gmail.com

Additional material to this book can be downloded from http://extras.springer.com

ISBN-10 3-642-42156-3 Springer Berlin Heidelberg New York
ISBN-13 978-3-642-42156-3 Springer Berlin Heidelberg New York

Springer is a part of Springer Science+Business Media
springeronline.com

Softcover re-print of the Hardcover 1st edition 2006

Cover design: E. Kirchner, Heidelberg
Production: Almas Schimmel
Typesetting: camera-ready by Author
Printing: Krips bv, Meppel
Binding: Stürtz AG, Würzburg

Printed on acid-free paper 30/3141/as 5 4 3 2 1 0

Sathyajith Mathew

Wind Energy

Fundamentals, Resource Analysis and Economics

Who has gathered the wind in his fists?

....***Proverbs 30:4***

Dedicated to my parents,
wife Geeta Susan and kids Manuel & Ann

Preface

Growing energy demand and environmental consciousness have re-evoked human interest in wind energy. As a result, wind is the fastest growing energy source in the world today. Policy frame works and action plans have already been formulated at various corners for meeting at least 20 per cent of the global energy demand with new-renewables by 2010, among which wind is going to be the major player.

In view of the rapid growth of wind industry, Universities, all around the world, have given due emphasis to wind energy technology in their undergraduate and graduate curriculum. These academic programmes attract students from diversified backgrounds, ranging from social science to engineering and technology. Fundamentals of wind energy conversion, which is discussed in the preliminary chapters of this book, have these students as the target group. Advanced resource analysis tools derived and applied are beneficial to academics and researchers working in this area. The Wind Energy Resource Analysis (WERA) software, provided with the book, is an effective tool for wind energy practitioners for assessing the energy potential and simulating turbine performance at prospective sites.

The introductory chapter narrates the historic development of wind energy technology along with its present status and future prospects. This is followed by Chapter 2, which presents the basic principles of wind energy conversion. Descriptions on different types of wind machines and their performances are briefed here. Basics of wind rotor aerodynamics and its application in the turbine design are also presented in this chapter.

The third chapter is devoted to the methods of measurement and analysis of wind spectra for energy use. Statistical methods for wind energy analysis are introduced here. These are further extended for developing models for estimating the wind energy potential of a prospective site.

Constructional features of various systems and sub-systems of a Wind Energy Conversion System (WECS) are described in Chapter 4. Along with wind electric generators, wind powered water pumping systems are also considered. Features of wind farms, both onshore and offshore, are also discussed in this chapter.

Chapter 5 deals with performance models of WECS. Tools to simulate the field performance of wind powered generators and water pumps are presented in this section. Optimal matching of WECS with the site is also discussed.

Sixth chapter is devoted to the environmental aspects of wind energy conversion. While highlighting the environment related merits of wind energy, the recent concerns over avian issues, visual impacts, noise etc. are not overlooked. A life cycle based approach is adopted for these discussions.

Economics of wind energy conversion is analysed in Chapter 7, following the present worth method. Factors affecting the costs and benefits of wind generated electricity are discussed and indices for economic appraisal are evolved.

Wind Energy Resource Analysis (WERA) software, which comes along with the book, is beneficial to readers who are not familiar with the numerical techniques applied in wind resource analysis. Illustrative examples included in all the chapters compliment the concepts presented in the text.

Subjects presented in this book are primarily derived from my experiences in teaching undergraduate and graduate engineering students. Research and field experiences on WECS have also helped me in formulating the materials presented. Further, serving as a resource person for various wind energy training programmes has also helped me in adopting a multi-disciplinary approach, which is essential for tackling a subject like wind energy. Hence, I would like to thank my students for their contribution.

Compiling information from various sources is essential for developing a book of this nature. I thank the authors of research papers and reports, which are referred in various chapters of this book. Several industries and organizations have supported me by providing information and materials which were essential for this project. Special thanks are due to Hawaiian Electric Company, Renewable Energy Systems Ltd, THALES instruments GmbH, Vaisala Oyj, Siemens Wind Power A/S, ReSoft, and Wikipedia, on this account.

I am fortunate to have the wholehearted support from my professors and colleagues for this project. Let me thank Prof. K.I. Koshy, Prof. C.P. Muhammad and Prof. Jippu Jacob for perusing the manuscript. Contribution of Prof. Anilkumar V and Dr. Damodar Rao in developing WERA is thankfully acknowledged. Thanks are also due to Prof. John D Burton, Prof. K.P Pandey, Prof. Ashok Alex Philip, Prof. Vishnu B, Dr. Dhalin D and Er. Nisha T.V, for their helps at various stages of this work.

As 'to err is human', suggestions for improving the content of this book in future are most welcome.

Sathyajith Mathew

Contents

1 Introduction

"Of all the forces of nature, I should think the wind contains the greatest amount of power"

........Abraham Lincoln

Energy is one of the crucial inputs for socio-economic development. The rate at which energy is being consumed by a nation often reflects the level of prosperity that it could achieve. Social and economic well being can be gauged by the Human Development Index (HDI), developed under the United Nations Development Programme (UNDP). It is found that most of the developed nations, showing high HDI, have per capita energy consumption in the range of 4000 to 9000 kilograms of oil equitant per annum [14]. On the other hand, the developing nations with lower per capita energy use (below 500 kgoe for most of the countries) could achieve HDI only below 0.5. For an HDI higher than 0.8, the per capita energy use has to be enhanced, at least to a level of 1000 kgoe.

Global population is increasing day by day. The population growth is more rapid in developing countries than the industrialized nations [4]. As a result of this population growth and developmental activities, the energy demand is also increasing. During the past 10 years, the primary energy use in the industrialized countries increased at a rate of 1.5 per cent per annum. The corresponding change in developing nations was 3.2 per cent [14]. With this trend prevailing, the global energy demand would increase considerably in the coming years. Future projections indicate that the Total Primary Energy Supply (TPES) should be increased to12,100 Mtoe by 2010 and 16, 300 Mtoe by 2030 [5].

The global energy demand is met from a variety of sources. Fossil fuels consisting of coal, oil, and natural gas meet around 80 per cent of the needs [15]. Share of nuclear power is approximately 7 per cent. The renewables supply 13.7 per cent to which traditional bio-mass and large hydro contribute the major share. At present, share of new renewables (wind, solar, etc.) is only 2.2 per cent. Hence, if the current scenario continues, we have to rely heavily on the fossil fuels to meet our energy needs.

Unfortunately, these fossil fuels are finite resources and will be completely exhausted one day or the other. The proved reserves of coal are only 566 Gtoe. Similarly, the reserves of oil and natural gas are limited to 143 Gtoe and 138 Gtoe respectively. Even at the current consumption rate of 2.26 Gtoe per annum, the proven coal reserve is sufficient for only the next 250 years.

Reserves of oil and natural gas also face similar situation. Though we could discover new reserves of these resources, the rate of discovery has been declining for the past 40 years. Hence, while our energy demand is increasing day by day, the available resources are depleting. This will definitely lead us to the much discussed energy crisis. However, the crisis may not be an imminent reality as the time scale may prolong due to discoveries of new resources.

Environmental ill effects of fossil fuel based power plants add another dimension to this problem. These power plants load the atmosphere with greenhouse gases (GHG) and particulates, resulting in global warming and climate change. Generation and consumption of energy is responsible for 50 to 60 per cent of GHG released into the atmosphere on a global basis. With the increase in energy use, atmospheric pollution from the energy sector is expected to increase further in the near future. Several deliberations are being made globally to chalk out means and measures to reduce the level of atmospheric pollution due to human activities. The Kyoto Protocol and the Johannesburg summit are considered as few positive steps towards a carbon free world. It is widely agreed that, to reduce the emission levels, at least 10 per cent of our energy supply should come from renewable sources.

Here comes the significance of sustainable energy sources like wind. The quantum of energy, associated with the wind is enormous. With today's technology, wind is an environment friendly and economically viable source of energy, which can be tapped in a commercial scale.

History of wind energy conversion along with its present status and future prospects are discussed briefly in the following sections.

1.1 History of wind energy

Human efforts to harness wind for energy date back to the ancient times, when he used sails to propel ships and boats. Later, wind energy served the mankind by energising his grain grinding mills and water pumps. During its transformation from these crude and heavy devices to today's efficient and sophisticated machines, the technology went through various phases of development.

There is disagreement on the origin of the concept of using wind for mechanical power. Some believe that the concept originated in ancient Babylonia. The Babylonian emperor Hammurabi planned to use wind power for his ambitious irrigation project during seventeenth century B.C. [3]. Others argue that the birth place of wind mills is India. In Arthasastra, a classic work in Sanskrit written by Kautiliya during 4th century B.C., references are seen on lifting water with contrivances operated by wind [12]. However, there are no records to prove that these concepts got transformed to real hardware.

The earliest documented design of wind mill dates back to 200 B.C. The Persians used wind mills for grinding grains during this period. Those were vertical axis machines having sails made with bundles of reeds or wood. The grinding stone was attached to the vertical shaft. The sails were attached to the central

Fig.1.1. An ancient windmill in the British Isles (Author: Michael Reeve, source: Wikipedia, http://wikipedia.org. GNU Free Documentation License applies to this image)

shaft using horizontal struts. The size of the sails was decided by the materials used for its fabrication, usually 5 m long and 9 m tall.

By the 13th century, grain grinding mills were popular in most of Europe. The French adopted this technology by 1105 A.D. and the English by 1191 A.D. In contrast with the vertical axis Persian design, European mills had horizontal axis. These post mills were built with beautiful structures. The tower was circular or polygonal in cross-section and constructed in wood or brick. The rotor was manually oriented to the wind by adjusting the tail. The mill was protected against high winds by turning the rotor out of the wind or removing the canvas covering the rotor.

The Dutch, with renowned designer Jan Adriaenszoon, were the pioneers in making these mills. They made many improvements in the design and invented several types of mills. Examples are the *tjasker* and *smock mills*. The rotors were made with crude airfoil profile to improve the efficiency. Apart from grain grinding, wind mills were employed to drain marshy lands in Holland. These wind mills reached America by mid-1700, through the Dutch settlers.

Fig.1.2. An ancient Spanish 'wind farm' (Author: Lourdes Cardenal, source: Wikipedia, http://wikipedia.org. GNU Free Documentation License applies to this image)

This is followed by the water pumping wind mill, which is still considered as one of the most successful application of wind power. The so-called American multi bladed wind turbine appeared in the wind energy history by the mid-1800. Relatively smaller rotors, ranging from one to several meters in diameter, were used for this application. The primary motive was to pump water from a few meters below the surface for agricultural uses. These *water pumpers*, with its metallic blades and better engineering design, offered good field performance. Over six million of such units were installed in US alone, between 1850 and 1930.

The era of wind electric generators began close to 1900's. The first modern wind turbine, specifically designed for electricity generation, was constructed in Denmark in 1890. It supplied electricity to the rural areas. During the same period, a large wind electric generator having 17 m 'picket fence' rotor was built in Cleveland, Ohio. For the first time, a speed-up gear box was introduced in the design. This system operated for 20 years generating its rated power of 12 kW.

More systematic methods were adopted for the engineering design of turbines during this period. With low-solidity rotors and aerodynamically designed blades, these systems could give impressive field performance. By 1910, several hundreds of such machines were supplying electrical power to the villages in Denmark. By about 1925, wind electric generators became commercially available in the American market. Similarly, two and three bladed propeller turbines ranging from 0.2 to 3 kW in capacity were available for charging batteries.

Turbines with bigger capacity were also developed during this period. The first utility-scale system was installed in Russia in 1931. A 100 kW turbine was installed on the Caspian sea shore, which worked for two years and generated about 20,000 kW electricity. Experimental wind plants were subsequently constructed in other countries like United States, Denmark, France, Germany, and Great Britain. A significant development in large-scale systems was the 1250 kW turbine fabricated by Palmer C. Putman. The turbine was put to use in 1941 at the Grandpa's Knob, near Rutland, Vermont [8]. Its 53 m rotor was mounted on a 34 m tall tower. This machine could achieve a constant rotor speed by changing the blade pitch. The machine operated for 1100 hours during the next five years, i.e., till the blades failed in 1945. The project is considered to be a succuss as it could demonstrate the technical feasibility of large- scale wind-electric generation.

Some interesting designs of wind turbine were experimented during this period. Darrieus G.J.M, a French engineer, putforth the design of Darrieus turbine in 1920, which was patented in United Sates in 1931 [9]. In contrast with the popular horizontal axis rotors, Darrieus turbines had narrow curved blades rotating about its vertical axis. During the same period, Julius D. Madaras invented a turbine working on Magnus effect. Magnus effect is basically derived from the force on a spinning cylinder placed in a stream of air. Another significant development at this time was the Savonius rotor in Finland, invented by S.J. Savonius. This rotor was made with two halves of a cylinder split longitudinally and arranged radially on a vertical shaft. The transverse cross-section of the rotor resembled the letter 'S' [10]. The rotor was driven by the difference in drag forces acting on its concave and convex halves, facing the wind.

Intensive research on the behaviour of wind turbines occurred during 1950's. The concept of high tip speed ratio-low solidity turbines got introduced during this period. For example, light-weight constant-speed rotors were developed in Germany in 1968. They had fibre glass blades attached to simple hollow towers supported by guy ropes. The largest of this breed was of 15 m diameter with a rated output of 100 kW.

In the later years, cheaper and more reliable electricity, generated from fossil fuel based plants became available. When the electricity generated from wind costed 12 to 30 cents/kWh in 1940, the same generated from other sources was available at 3 to 6 cents/kWh [7]. Cost of electricity from fossil fuels further declined below 3 cents/kWh by 1970. Fossil fuels were available in plenty at a relatively cheaper rate at that time. Several nuclear power projects were also embarked on, believing that it would be the ultimate source for the future energy needs. Thus, the interest in wind energy declined gradually, especially by 1970.

The oil crisis in 1973, however, forced the scientists, engineers and policy makers to have a second thought on the fossil fuel dependence. They realised that political tampering can restrict the availability and escalate the cost of fossil fuels. Moreover, it was realised that the fossil fuel reserve would be exhausted one day or the other. Nuclear power was unacceptable to many, due to safety reasons. These factors caused the revival of interest in wind energy. Research on resource analysis, hardware development, and cost reduction techniques were intensified. United States entrusted its National Aeronautics and Space Administration (NASA) with the development of

1.3. The MOD OA wind turbine (Courtesy of Hawaiian Electric Company, Inc., http://heco.com)

large wind turbines. As a result, a series of horizontal axis turbines named MOD-0, MOD-1, MOD-2 and MOD-5 were developed [6]. These projects were stopped by mid-1980's due to various reasons. During the same period, scientists at Sandia Laboratories focussed their research on the design and development of the Darrieus turbine [11]. They fabricated several models of the Darrieus machine in different sizes during 1980's.

Research and development on wind energy are seen intensified in the later years. A few innovative concepts like the vortex turbine, diffuser augmented design, Musgrove rotor etc. were also proposed during that time. Prototypes of these turbines were constructed and tested. However, only the horizontal axis propeller design could emerge successfully on a commercial scale.

1.2 Current status and future prospects

Owing to our commitment to reduce GHG emissions and provide adequate energy to the developing world, efforts are being made to supplement our energy base with renewable sources. Several countries have already formulated policy frameworks to ensure that renewables play an impressive role in the future energy scenario. For example, the European Union targets to meet 22 per cent of their

1.4 A wind farm (Hamish Hill/Courtesy of Renewable Energy Systems Ltd., www.res-ltd.com)

demand from renewables by 2010. Wind, being the commercially viable and economically competitive renewable source, is going to be the major player in meeting this target.

Wind is the world's fastest growing energy source today and it has been retaining this position consecutively for the last five years. The global wind power capacity has increased by a factor of 4.2 during the last five years. The total global installed capacity is 39434 MW in 2004. Installed capacity in different regions is shown in Fig. 1.5. Over 73 per cent of the global installations are in Europe. Germany is the European leader, followed by Spain and Denmark. The five countries, leading in wind energy generation are listed in Table 1.1.

With the increasing thrust on renewables and reducing cost of wind generated electricity, the growth of wind energy will continue in the years to come. According to European Wind Energy Association (EWEA), wind with its expected 230,000 MW installation, can supply 12 per cent of the global energy demand by 2010 [1]. This indicates a market worth around 25 billion Euros. The installed capacity may reach a level of 1.2 million MW by 2020.

Table 1.1 Global leaders in wind energy generation

Country	Installed capacity, MW
Germany	14609
United States of America	6352
Spain	6202
Denmark	3115
India	2120

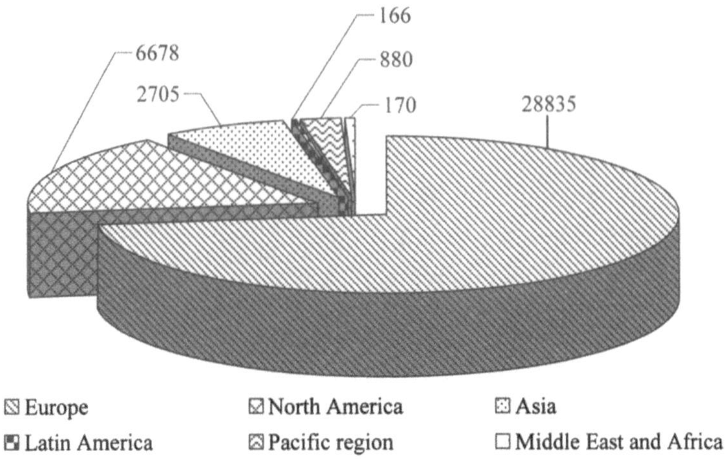

Fig. 1.5. Installed Wind energy capacity (MW) in different regions [13]

In tune with the growth of the industry, the wind energy technology is also changing. One apparent change is the shift towards offshore installations. Several ambitious offshore projects are in the pipeline. For example, 20 offshore projects are planned to be installed in UK by 2006, with a total capacity of 1400 MW [16]. In Germany, around 30 offshore projects worth 60,000 MW are in various stages of processing. In United States also, the offshore activities are intensifying.

Another trend in the industry is to go for larger machines. As bigger turbines are cheaper on a unit kW basis, the industry is growing from MW to multi-MW scale. The 2 MW+ sector is rapidly growing. Several manufactures like the RE Power Systems AG are coming up with turbines of even 5 MW size. The RE Power model is equipped with a huge 125 m rotor having each blade weighing around 19 tonnes [2]. Efforts are also on to reduce the total head mass (THM) which is the total mass of nacelle and rotor. Reduction in THM has positive impact on system dynamics. By clever engineering design, NEG Micon could restrict the THM of their 4.2 MW model to 214 tonnes, which is a remarkable achievement. Owing to the active grid support and better efficiency, the variable speed option with double fed induction generator is getting more prominence in the industry. Another innovative concept that may prove effective in future is the direct drive machines.

References

1. de Azua CR, Colasimone L (2003) Record growth for global wind power in 2002; 28% increase, wind technology worth $7.3 billion installed last year. AWEA-EWEA News release, Global Wind Power Installations, http://www.ewea.org
2. de Vries E (2003) Wind turbine technology trends – review 2003. Renewable Energy World 6(4): 154-167

3. Golding E (1976) The generation of electricity by wind power. Halsted Press, New York
4. International Energy Agency (2003) Energy balances of non-OECD Countries 2000-2001, Paris : IEA and OECD
5. International Energy Agency (2003) Key world energy statistics. France, http://www.iea.org
6. Johnson GL (2001) Wind energy systems. http://www.rpc.com.au
7. Kloeffler RG, Sitz EL (1946) Electric energy from winds. Kansas State College of Engineering Experiment Station Bulletin 52, Manhattan, Kans
8. Putnam PC (1948) Power from the wind. Van Nostrand, New York
9. Ramler JR, Donovan RM (1979) Wind turbines for electric utilities: Development status and economics. DOE/NASA/1028-79/23, NASA TM-79170, AIAA-79-0965
10. Savonius SJ (1931) The S-rotor and its applications. Mechanical Engineering 53(5) :333-338
11. Sheldahl RE, Blackwell BF (1977) Free-air performance tests of a 5-meter-diameter darrieus turbine. Sandia Laboratories Report SAND 77-1063
12. Sorensen B (1995) History of, and recent progress in, wind-energy utilization. Annual Review of Energy and the Environment 20(1) : 387-424
13. The Windicator (2005) Wind energy facts and figures from windpower monthly. Windpower Monthly News Magazine, Denmark, USA : 1-2
14. UNDP, World Energy Council (2004) World energy assessment: overview 2004 update. Bureau for development policy, New York : 25-31
15. World Energy Council (2000) World Energy Assessment: Energy and the challenge of sustainability. New York
16. Zaaijer M, Henderson A (2003) Offshore update – A global look at offshore wind energy. Renewable Energy World 6(4): 102-119

2 Basics of wind energy conversion

Energy available in wind is basically the kinetic energy of large masses of air moving over the earth's surface. Blades of the wind turbine receive this kinetic energy, which is then transformed to mechanical or electrical forms, depending on our end use. The efficiency of converting wind to other useful energy forms greatly depends on the efficiency with which the rotor interacts with the wind stream. In this chapter, let us discuss the fundamental principles involved in this wind energy conversion process.

2.1 Power available in the wind spectra

The kinetic energy of a stream of air with mass m and moving with a velocity V is given by

$$E = \frac{1}{2} m V^2 \tag{2.1}$$

Consider a wind rotor of cross sectional area A exposed to this wind stream as shown in Fig. 2.1. The kinetic energy of the air stream available for the turbine can be expressed as

$$E = \frac{1}{2} \rho_a v V^2 \tag{2.2}$$

where ρ_a is the density of air and v is the volume of air parcel available to the rotor. The air parcel interacting with the rotor per unit time has a cross-sectional area equal to that of the rotor (A_T) and thickness equal to the wind velocity (V). Hence energy per unit time, that is power, can be expressed as

$$P = \frac{1}{2} \rho_a A_T V^3 \tag{2.3}$$

From Eq. (2.3), we can see that the factors influencing the power available in the wind stream are the air density, area of the wind rotor and the wind velocity. Effect of the wind velocity is more prominent owing to its cubic relationship with the power.

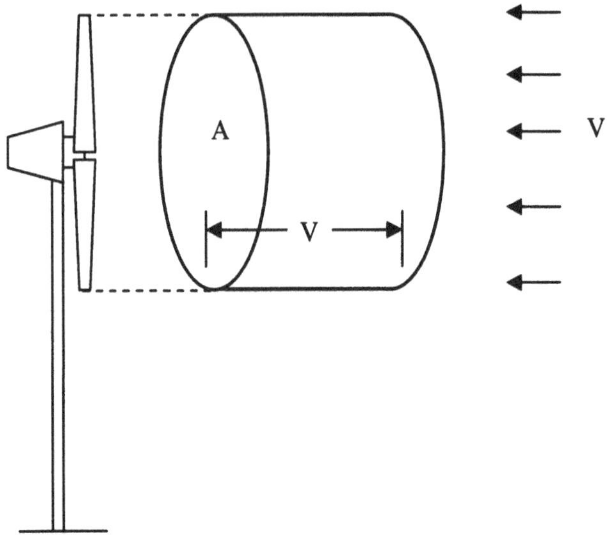

Fig. 2.1. An air parcel moving towards a wind turbine

Factors like temperature, atmospheric pressure, elevation and air constituents affect the density of air. Dry air can be considered as an ideal gas. According to the ideal gas law,

$$p\,V_G = nRT \tag{2.4}$$

where p is the pressure, V_G is the volume of the gas, n is the number of kilo moles of the gas, R is the universal gas constant and T is the temperature. Density of air, which is the ratio of the mass of 1 kilo mole of air to its volume, is given by

$$\rho_a = \frac{m}{V_G} \tag{2.5}$$

From Eqs. (2.4) and (2.5), density is given by

$$\rho_a = \frac{m\,p}{RT} \tag{2.6}$$

If we know the elevation Z and temperature T at a site, then the air density can be calculated by

$$\rho_a = \frac{353.049}{T}\, e^{\left(-\,0.034\,\frac{Z}{T}\right)} \tag{2.7}$$

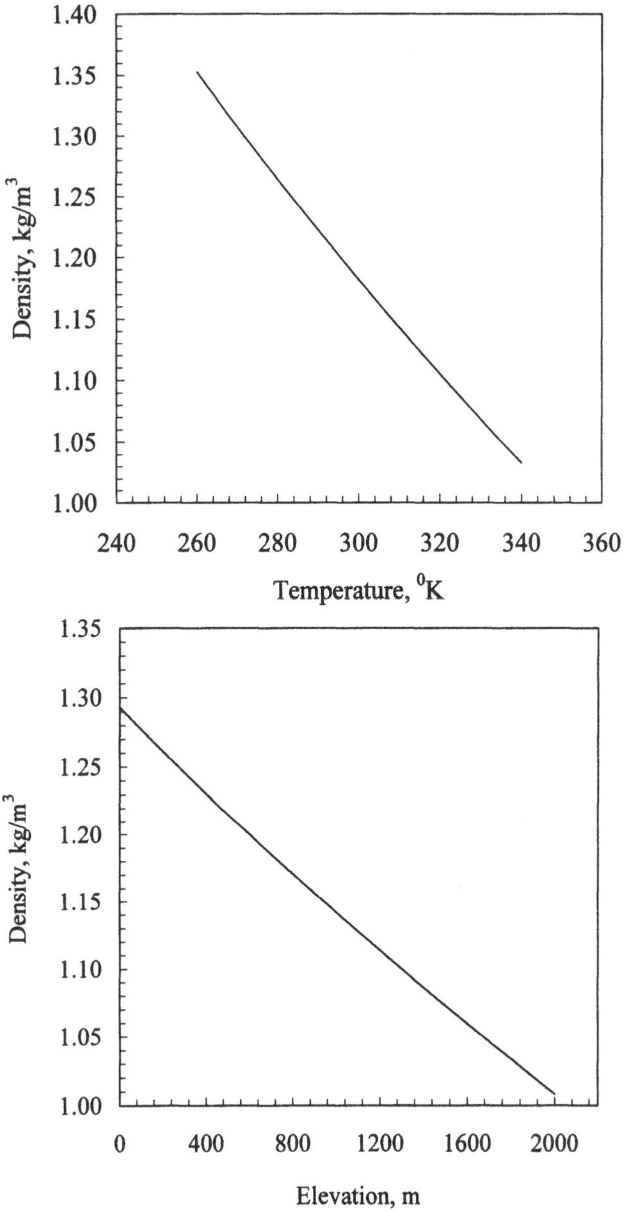

Fig. 2.2. Effect of elevation and temperature on air density

The density of air decreases with the increase in site elevation and temperature as illustrated in Fig. 2.2. The air density may be taken as 1.225 for most of the practical cases. Due to this relatively low density, wind is rather a diffused source of energy. Hence large sized systems are often required for substantial power production.

The most prominent factor deciding the power available in the wind spectra is its velocity. When the wind velocity is doubled, the available power increases by 8 times. In other words, for the same power, rotor area can be reduced by a factor of 8, if the system is placed at a site with double the wind velocity. The advantages are obvious. Hence, selecting the right site play a major role in the success of a wind power projects. We will discuss this aspect in detail in the coming chapters.

2.2 Wind turbine power and torque

Theoretical power available in a wind stream is given by Eq. (2.3). However, a turbine cannot extract this power completely from the wind. When the wind stream passes the turbine, a part of its kinetic energy is transferred to the rotor and the air leaving the turbine carries the rest away. Actual power produced by a rotor would thus be decided by the efficiency with which this energy transfer from wind to the rotor takes place. This efficiency is usually termed as the power coefficient (C_p). Thus, the power coefficient of the rotor can be defined as the ratio of actual power developed by the rotor to the theoretical power available in the wind. Hence,

$$C_p = \frac{2\, P_T}{\rho_a\, A_T\, V^3} \tag{2.8}$$

where P_T is the power developed by the turbine. The power coefficient of a turbine depends on many factors such as the profile of the rotor blades, blade arrangement and setting etc. A designer would try to fix these parameters at its optimum level so as to attain maximum C_p at a wide range of wind velocities.

The thrust force experienced by the rotor (F) can be expressed as

$$F = \frac{1}{2}\, \rho_a\, A_T\, V^2 \tag{2.9}$$

Hence we can represent the rotor torque (T) as

$$T = \frac{1}{2}\, \rho_a\, A_T\, V^2 R \tag{2.10}$$

where R is the radius of the rotor. This is the maximum theoretical torque and in practice the rotor shaft can develop only a fraction of this maximum limit. The ratio between the actual torque developed by the rotor and the theoretical torque is termed as the torque coefficient (C_T). Thus, the torque coefficient is given by

$$C_T = \frac{2\, T_T}{\rho_a\, A_T\, V^2 R} \tag{2.11}$$

where T_T is the actual torque developed by the rotor.

The power developed by a rotor at a certain wind speed greatly depends on the relative velocity between the rotor tip and the wind. For example, consider a situation in which the rotor is rotating at a very low speed and the wind is approaching the rotor with a very high velocity. Under this condition, as the blades are moving slow, a portion of the air stream approaching the rotor may pass through it without interacting with the blades and thus without energy transfer. Similarly if the rotor is rotating fast and the wind velocity is low, the wind stream may be deflected from the turbine and the energy may be lost due to turbulence and vortex shedding. In both the above cases, the interaction between the rotor and the wind stream is not efficient and thus would result in poor power coefficient.

The ratio between the velocity of the rotor tip and the wind velocity is termed as the tip speed ratio (λ). Thus,

$$\lambda = \frac{R\,\Omega}{V} = \frac{2\pi N R}{V} \tag{2.12}$$

where Ω is the angular velocity and N is the rotational speed of the rotor. The power coefficient and torque coefficient of a rotor vary with the tip speed ratio. There is an optimum λ for a given rotor at which the energy transfer is most efficient and thus the power coefficient is the maximum ($C_{P\,max}$).

Now, let us consider the relationship between the power coefficient and the tip speed ratio.

$$C_p = \frac{2\,P_T}{\rho_a\,A_T\,V^3} = \frac{2\,T_T\,\Omega}{\rho_a\,A_T\,V^3} \tag{2.13}$$

Dividing Eq. (2.13) by Eq. (2.11) we get

$$\frac{C_p}{C_T} = \frac{R\,\Omega}{V} = \lambda \tag{2.14}$$

Thus, the tip speed ratio is given by the ratio between the power coefficient and torque coefficient of the rotor.

Example

Consider a wind turbine with 5 m diameter rotor. Speed of the rotor at 10 m/s wind velocity is 130 r/min and its power coefficient at this point is 0.35. Calculate the tip speed ratio and torque coefficient of the turbine. What will be the torque available at the rotor shaft? Assume the density of air to be 1.24 kg/m^3.

Area of the rotor is

$$A_T = \frac{\pi}{4} \times 5^2 = 19.63\,\text{m}^2$$

As the speed of the rotor is 130 r/min, its angular velocity is

$$\Omega = \frac{2\times\pi\times 130}{60} = 13.6 \text{ rad/s}$$

The tip speed ratio at this velocity is

$$\lambda = \frac{2.5 \times 13.6}{10} = 3.4$$

The torque coefficient is

$$C_T = \frac{0.35}{3.4} = 0.103$$

From this, torque developed can be calculated as

$$T_T = \frac{1}{2} \times 1.24 \times \frac{\pi}{4} \times 5^2 \times 10^2 \times 0.103 = 313.39 \text{ Nm}$$

2.3 Classification of wind turbines

Fig. 2.3. Three bladed horizontal axis wind turbines (Hamish Hill/Courtesy of Renewable Energy Systems Ltd., www.res-ltd.com)

Since the inception of the wind energy technology, machines of several types and shapes were designed and developed around different parts of the world. Some of these are innovative designs which are not commercially accepted. Although there are several ways to categorize wind turbines, they are broadly classified into horizontal axis machines and vertical axis machines, based on their axis of rotation.

2.3.1 Horizontal axis wind turbines

Horizontal axis wind turbines (HAWT) have their axis of rotation horizontal to the ground and almost parallel to the wind stream (Fig. 2.3). Most of the commercial wind turbines fall under this category. Horizontal axis machines have some distinct advantages such as low cut-in wind speed and easy furling. In general, they show relatively high power coefficient. However, the generator and gearbox of these turbines are to be placed over the tower which makes its design more complex and expensive. Another disadvantage is the need for the tail or yaw drive to orient the turbine towards wind.

Depending on the number of blades, horizontal axis wind turbines are further classified as single bladed, two bladed, three bladed and multi bladed, as shown in Fig. 2.4. Single bladed turbines are cheaper due to savings on blade materials. The drag losses are also minimum for these turbines. However, to balance the blade, a counter weight has to be placed opposite to the hub. Single bladed designs are not very popular due to problems in balancing and visual acceptability. Two bladed rotors also have these drawbacks, but to a lesser extent. Most of the present commercial turbines used for electricity generation have three blades.

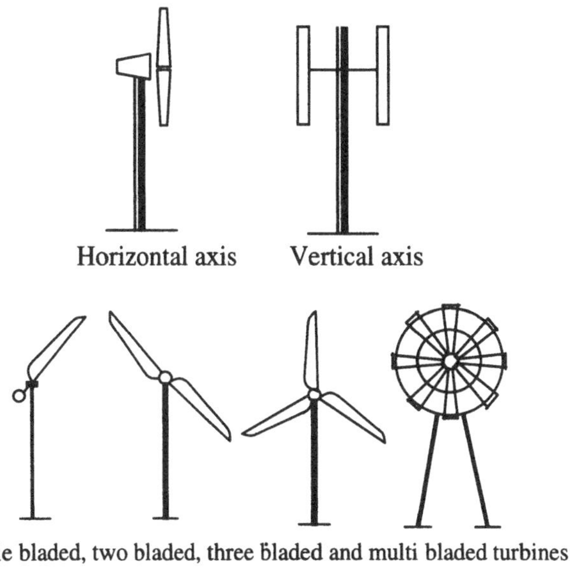

Single bladed, two bladed, three bladed and multi bladed turbines

Fig. 2.4. Classification of wind turbines

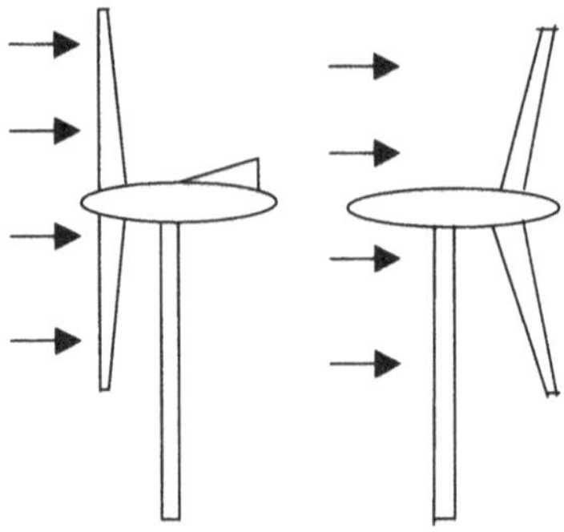

Fig. 2.5. Upwind and downwind turbines

They are more stable as the aerodynamic loading will be relatively uniform. Machines with more number of blades (6, 8, 12, 18 or even more) are also available. The ratio between the actual blade area to the swept area of a rotor is termed as the solidity. Hence, multi-bladed rotors are also called high solidity rotors. These rotors can start easily as more rotor area interacts with the wind initially. Some low solidity designs may require external starting.

Now consider two rotors, both of the same diameter, but different in number of blades; say one with 3 blades and the other with 12 blades. Which will produce more power at the same wind velocity? As the rotor swept area and velocity are the same, theoretically both the rotors should produce the same power. However aerodynamic losses are more for the rotor with more number of blades. Hence, for the same rotor size and wind velocity, we can expect more power from the three bladed rotor.

Then why do we need turbines with more blades? Some applications like water pumping require high starting torque. For such systems, the torque required for starting goes up to 3-4 times the running torque. Starting torque increases with the solidity. Hence to develop high starting torque, water pumping wind mills are made with multi bladed rotors.

Based on the direction of receiving the wind, HAWT can be classified as upwind and down wind turbines as shown in Fig. 2.5. Upwind turbines have their rotors facing the wind directly. As the wind stream passes the rotor first, they do not have the problem of tower shadow. However, yaw mechanism is essential for such designs to keep the rotor always facing the wind. On the other hand, downwind machines are more flexible and may not require a yaw mechanism. But, as the rotors are placed at the lee side of the tower, there may be uneven loading on the blades as it passes through the shadow of the tower.

2.3.2 Vertical axis wind turbines

The axis of rotation of vertical axis wind turbine (VAWT) is vertical to the ground and almost perpendicular to the wind direction as seen from Fig. 2. 4. The VAWT

Fig. 2.6. Darrieus wind turbine (Source: Wikipedia, http://wikipedia.org. GNU Free Documentation License applies to this image)

can receive wind from any direction. Hence complicated yaw devices can be eliminated. The generator and the gearbox of such systems can be housed at the ground level, which makes the tower design simple and more economical. Moreover the maintenance of these turbines can be done at the ground level. For these systems, pitch control is not required when used for synchronous applications.

The major disadvantage of some VAWT is that they are usually not self starting. Additional mechanisms may be required to 'push' and start the turbine, once it is stopped. As the rotor completes its rotation, the blades have to pass through aerodynamically dead zones which will result in lowering the system efficiency. There are chances that the blades may run at dangerously high speeds causing the system to fail, if not controlled properly. Further, guy wires are required to support the tower structure which may pose practical difficulties. Features of some major vertical axis designs are discussed below.

Darrieus rotor

Darrieus rotor, named after its inventor Georges Jeans Darrieus, works due to the lift force generated from a set of airfoils (Fig. 2.6). In the original design the blades are shaped like egg beaters or *troposkein* (turning rope) and are under pure tension while in operation. This typical blade configuration helps in minimizing the bending stress experienced by the blades. There are several variations in the Darrieus design of which some are with straight vertical blades, usually called

Fig. 2.7. A low cost Savonius wind turbine with rotors arranged 90° out of phase

Giromills. Darrieus rotor usually works at high tip speed ratio which makes it attractive for wind electric generators. However, they are not self-starting and require external 'excitation' to cut-in. Moreover, the rotor produces peak torque only twice per revolution.

Savonius rotor

The Savonius wind turbine, invented by S.J. Savonius, is a vertical axis machine consisting of two half cylindrical (or elliptical) blades arranged in 'S' shape (Fig. 2.7). Convex side of one of the half cylinder and the concave side of the other are facing the wind at a time as shown in Fig. 2.8. The basic driving force of Savonius rotor is drag. The drag coefficient of a concave surface is more than the convex surface. Hence, the half cylinder with concave side facing the wind will experience more drag force than the other cylinder, thus forcing the rotor to rotate. Some times two or more rotors fixed one over the other at 90^{0} offset may be used to smoothen the torque fluctuations during rotation. Another way to improve the performance is to attach deflector augmenters with the rotor. The augmenter shades the convex half facing the wind and directs the flow to the concave half thus enhancing the performance.

Being drag machines, Savonius rotors have relatively lower power coefficient. However, some experimental rotors have shown power coefficient upto 35 per cent. These rotors have high solidity and thus high starting torque.

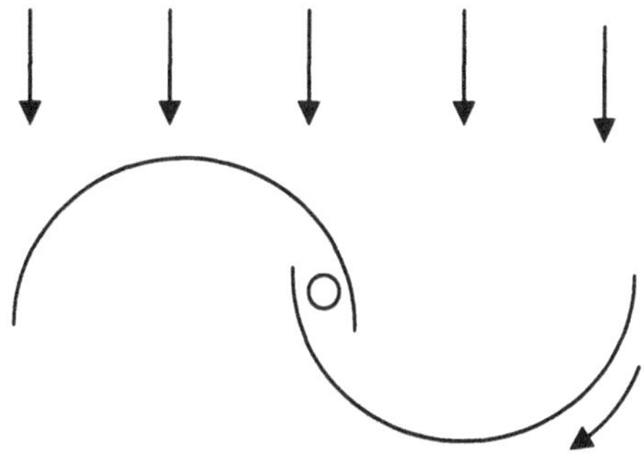

Fig. 2.8. Principle of Savonius rotor

They work at low tip speed ratios, with the maximum of about 1. They are very simple in construction-even can be made from oil barrels cut in two halves lengthwise. Hence they are preferred for high torque-low speed applications like water pumping.

Musgrove rotor

Musgrove rotor was developed by a research team under Prof. Musgrove at the Reading University, UK. It is basically a vertical axis lift machine having 'H' shaped blades and a central shaft (Fig. 2.9). At high wind speeds the rotor feathers and turn about a horizontal point due to the centrifugal force. This eliminates the risk of higher aerodynamic forces on the blades and the structure.

Based on driving aerodynamic force, wind turbines are classified as lift and drag machines. Turbines that work predominantly by the lift force are called the lift machines and that by the drag force are called drag turbines. It is always advantageous to utilize the lift force to run the turbine. Wind turbines are available in various sizes ranging from a fraction of kW to several MW. Based on the size, the turbines may be classified as small (< 25 kW) medium (25-100 kW), large (100-1000 kW) and very large (>1000 kW) machines.

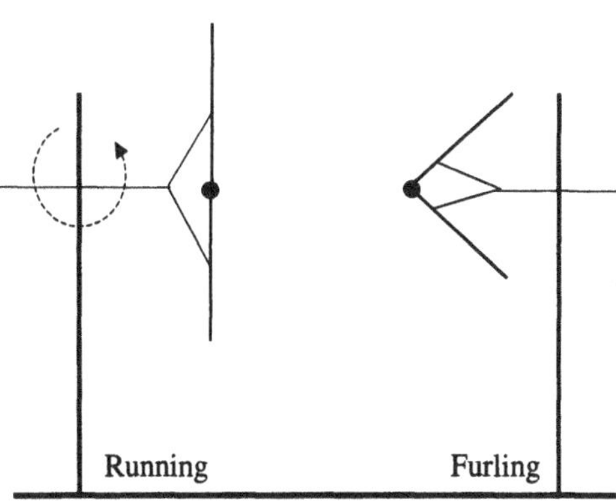

Fig. 2.9. Principle of Musgrove turbine

2.4 Characteristics of wind rotors

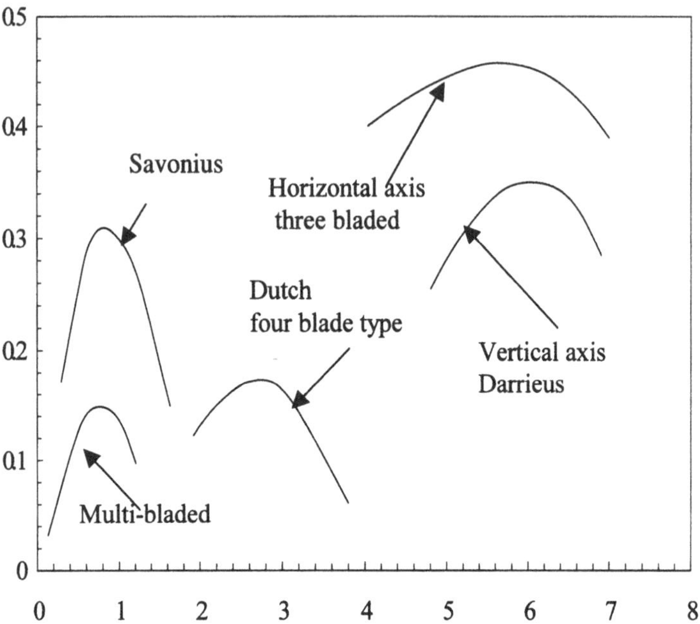

Fig. 2.10. Performance characteristics of wind rotors

The efficiency with which a rotor can extract power from the wind depends on the dynamic matching between the rotor and the wind stream. Hence, the performance of a wind rotor is usually characterised by the variations in its power coefficient with the tip speed ratio. As both these parameters are dimensionless, the C_P-λ curve will represent the rotor performance irrespective of the rotor size and site parameters. Once such a relationship could be deduced for a typical rotor design, it can further be translated to the velocity power curve of the rotor for practical applications.

Typical C_P-λ curves for different rotors are presented in Fig. 2.10. In general, initially the power coefficient of the turbine increases with the tip speed ratio, reaches a maximum at a typical λ, and then decreases with further increase in the tip speed ratio. These variations in C_P with λ depend on several design features of the rotor. American multi-bladed rotors show the lowest power coefficient and work at low speed ratio with the wind. A typical value for its peak power coefficient may be 14 per cent at a tip speed ratio of 0.8. However they have high solidity and hence high starting torque which make them attractive for water pumping. Two and three bladed propeller turbines and the Darrieus design work at higher tip speed ratios and show better efficiency. Hence they are suitable for wind-electric generators.

Savonius rotors, with its high solidity, work at lower tip speed ratio. Although, theoretically it is shown that the peak efficiency of such rotors cannot cross the limit of 20 per cent, Savonius reported a peak efficiency of 31 per cent under the wind tunnel test and 37 per cent in the free air stream [11]. Efficiencies ranging from 25 to 35 per cent are reported under various investigations on this rotor [6,8]. These values are quite impressive as these rotors are easier to fabricate and cost effective.

What is the maximum theoretical efficiency that a designer can expect from his wind turbine? Albert Betz, a German physicist, in 1962 had established a limit for the maximum power coefficient for an ideal wind rotor. He applied the axial momentum theory in its simplest form for his analysis and established that the maximum theoretical power coefficient of a wind turbine, operated predominantly by lift force, is 16/27 (59.3 per cent). This is known as the Betz limit (we will discuss the axial momentum theory under aerodynamic analysis of wind turbines). On the other hand, the maximum expected power coefficient of a drag machine is 8/27. This is why lift machines are preferred over drag machines for wind energy conversion. It should be noted that these are the theoretical values and some drag turbines like the Savonius rotors could show higher efficiencies under field evaluation as mentioned earlier.

2.5 Aerodynamics of wind turbines

Aerodynamics deals with the motion of air or other gaseous fluids and the forces acting on bodies moving through them. During the course of wind turbine development, several efforts were made to understand and interpret the aerodynamic principles governing wind energy conversion and to apply them in the successful system design. Earlier initiatives in this direction relied more on the aviation industry. Aerodynamic theories developed for airplanes and helicopters were adopted for defining the performance of wind turbines. However, now, theories are specifically formulated for wind turbines which are further refined and reinforced with the help of experimental techniques. Let us discuss some of these basic principles in the following sections.

2.5.1 Airfoil

For the efficient energy extraction, blades of modern wind turbine are made with airfoil sections. Major features of such an airfoil are shown in Fig. 2.11. The airfoils used for the earlier day's wind turbines were the aviation air foils under the NACA (National Advisory Committee for Aeronautics) series. NACA specifies the features of the airfoil by numbers.

For example, in a four digit specification, the first number denotes the maximum camber of the airfoil at the chord line (in per cent of chord), the second number gives the location of the point of maximum camber from the leading edge

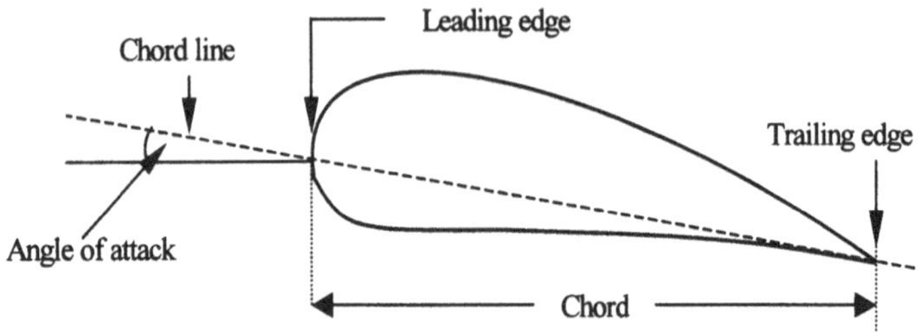

Fig 2.11. Important parameters of an airfoil

(in tenth of the chord) and the third and fourth numbers indicate the maximum thickness (in per cent of the chord). Thus a NACA 2415 air foil have maximum camber of 2 per cent, located at 0.4 times the chord length from the leading edge and the maximum thickness is 15 per cent of the chord.

The five digit specification of airfoil is similar, but information on the lift characteristics are also included. In a five-numbered NACA airfoil, the design lift coefficient in tenth is given by 1.5 times of the first digit. One-half of the distance from the leading edge to the location of maximum camber (in per cent of the chord) is represented by the second and third digits. The fourth and fifth digits are the thickness of the airfoil (in per cent of the chord). Thus a NACA 23012 series airfoil will have a design lift coefficient of 0.3. The maximum camber of the airfoil is 0.15 times the chord and its thickness is 12 per cent. Blades with higher digit NACA numbers are also available.

As far as the requirements of a wind turbine are concerned, NACA specified airfoils are poor in their stall characteristics. Moreover they are insensitive to wide variations in Reynolds number under which a wind machine is expected to work. Adding to these, they have low structural efficiency at the root region. These limitations called for the development of air foils tailored for wind turbines. Several research organizations like the Delft University of Technology [14],

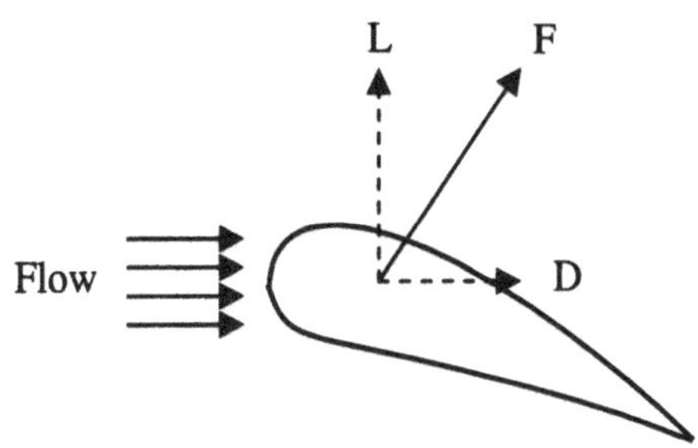

Fig. 2.12. Airfoil lift and drag

National Renewable Energy laboratory [13], and Riso National Laboratory [2] are engaged in research and development of airfoils exclusively for wind energy systems.

When an airfoil is placed in a wind stream, air passes through both upper and lower surfaces of the blade. Due to the typical curvature of the blade, air passing over the upper side has to travel more distance per unit time than that passing through the lower side. Thus the air particles at the upper layer move faster. According to the Bernoulli's theorem, this should create a low-pressure region at the top of the airfoil. This pressure difference between the upper and lower surfaces of the airfoil will result in a force F. The component of this force perpendicular to the direction of the undisturbed flow is called the lift force L (Fig. 2.12). The force in the direction of the undisturbed flow is called the drag force D.
The lift force is given by

$$L = C_L \frac{1}{2} \rho_a A V^2 \tag{2.15}$$

and the drag force (D) by

$$D = C_D \frac{1}{2} \rho_a A V^2 \tag{2.16}$$

where C_L and C_D are the lift and drag coefficients respectively.

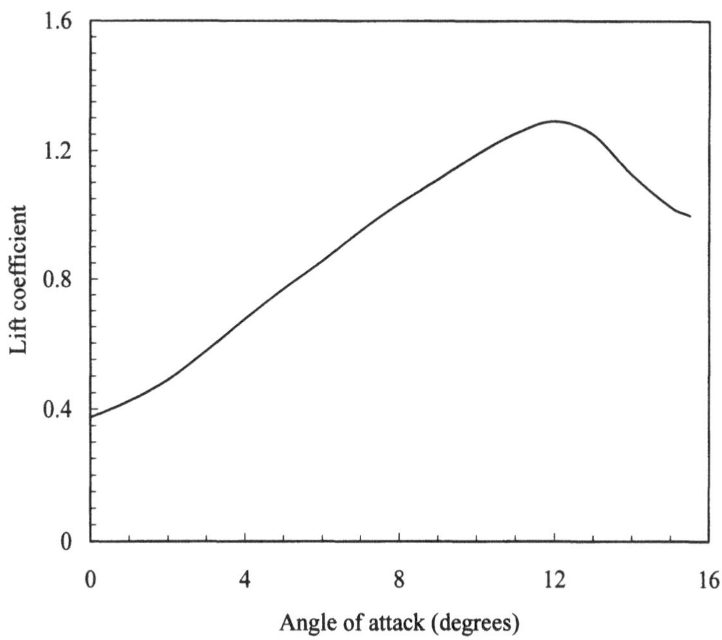

Fig. 2.13. Effect of angle of attack on airfoil lift

The angle between the undisturbed wind direction and the chord of the airfoil is known as the angle of attack (α). Lift and drag forces experienced by an airfoil is influenced by α. Fig. 2.13 illustrates the effect of angle of attack on the lift coefficient of an airfoil. At lower angles of attack, the lift force increases with α. The lift reaches its maximum at a certain α for an airfoil (12° in this example) and then decreases rapidly with further increase in α. This is because, at high angles of attack, the airflow enters an excessively turbulent region and the boundary layers get separated from the airfoil. At this region, lift force decreases and drag force is rapidly built up, resulting in the stall of the blade.

C_D-C_L characteristics of an airfoil can be established under wind tunnel experiments. Airfoils are fixed at different angles of attack in the wind tunnel flow and the forces of lift and drag acting on the foil are measured using transducers placed in the vertical and horizontal planes. From this corresponding lift and drag coefficients are calculated using Eqs. (2.15) and (2.16). In case of a wind turbine, our objective is to maximise the lift force and minimise the drag. Hence, in a given flow, it is very important to place our airfoil at an optimum angle of attack so that the C_D/C_L ratio is the minimum.

Consider the cross section of the rotating blade of a wind turbine as in Fig. 2.14. Apart from the wind velocity 'V', a point at the section is subjected to a velocity V_T due to the rotation of the rotor as shown in the figure. Thus the velocity V_R experienced at this point is the resultant of V and V_T. V_R will have lift and drag components as shown in the figure. Under this condition, the angle of attack α is the angle between 'V_R' and the chord line of the airfoil. For the same rotational speed, V_T at different sections of the blades varies with the distance of the section from the hub. Hence, the angle at which the resultant velocity approaches the rotor would also be different along the blade section, being steeper at the root of the blade. As we have seen, the C_D/C_L ratio for an airfoil is minimum at a particular angle of attack. To maintain this optimum attack angle throughout the blade sections, the blade may be twisted along its length.

Another factor affecting the lift and drag forces developed by an airfoil is the Reynolds number. Reynolds number is the ratio between the gravitational force and the viscous force.

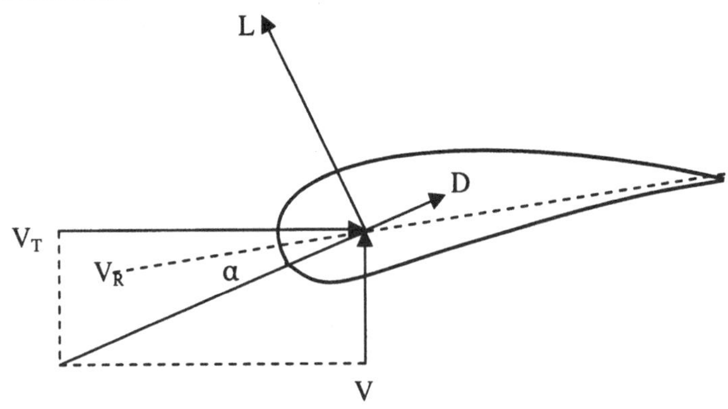

Fig. 2.14. Section of a rotating blade

Thus, considering the flow around an airfoil

$$R_e = \frac{V\ C}{\gamma} \tag{2.17}$$

where 'V' is the flow velocity, C is the chord length and γ is the kinematic viscosity of the fluid. For air, the kinematic viscosity at 20° C is 15×10^{-6} m²/s. For example, when an airfoil of chord 0.33 m is placed in a flow of 7 m/s, R_e of the flow is 1.54×10^5. However, in many preliminary calculations, the effect of R_e may be neglected as it has only a second order effect on the lift-drag characteristics.

2.5.2 Aerodynamic theories

Different theories are proposed to analyze the aerodynamics of wind turbines. These theories give an insight to the behavior of the rotor under varying operating conditions. Let us discuss some of the fundamental theories among them, applicable to HAWT.

Axial momentum theory

The conventional analysis of HAWT originates from the axial momentum concept introduced by Rankine, which was further improved by Froudes for marine propellers. Ideal flow conditions are considered for this analysis. The flow is assumed to be incompressible and homogeneous. The rotor is considered to be made up of infinite number of blades. Static pressures far in front and behind the rotor are considered to be equal to the atmospheric pressure. Frictional drag over the blades and wake behind the rotor are neglected.

Consider a wind turbine with rotor of area A_T, placed in a wind stream as shown in Fig. 2.15. Let A and $A^{'}$ be the areas of the sections 1-1, and 2-2 and V and $V^{'}$ are the respective wind velocities at these sections.

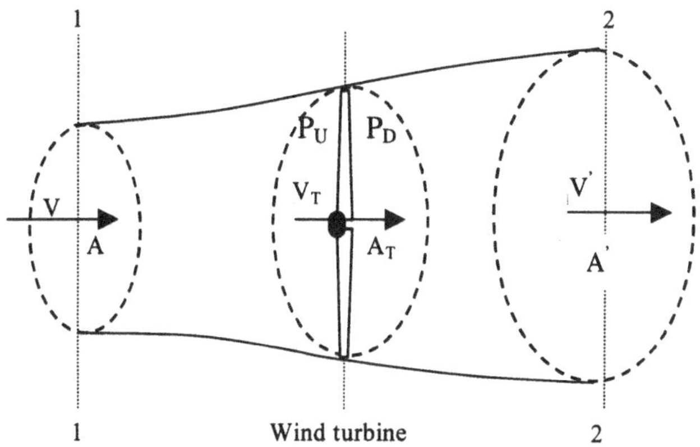

Fig. 2.15. The axial stream tube model

V_T is the velocity at the turbine section. According to the law of conservation of mass, the mass of air flowing through these sections is equal. Thus:

$$\rho_a A V = \rho_a A_T V_T = \rho_a A'V' \tag{2.18}$$

The thrust force experienced by the rotor is due to the difference in momentum of the incoming and outgoing wind, which is given by

$$F = \rho_a AV^2 - \rho_a A'V'^2 \tag{2.19}$$

As $AV = A'V' = A_T V_T$ from (2.18), the thrust can be expressed as

$$F = \rho_a A_T V_T \left(V - V'\right) \tag{2.20}$$

The thrust can also be represented as the pressure difference in the upstream and down stream sides of the rotor. Let p_U and p_D be the pressure at the upstream and down stream side of the rotor respectively. Hence:

$$F = \left(p_U - p_D\right) A_T \tag{2.21}$$

Applying the Bernoulli's equation at the sections and considering the assumption that the static pressures at sections 1-1 and 2-2 are equal to the atmospheric pressure p, we get

$$p + \frac{\rho_a V^2}{2} = p_U + \frac{\rho_a {V_T}^2}{2} \tag{2.22}$$

and

$$p + \frac{\rho_a V'^2}{2} = p_D + \frac{\rho_a {V_T}^2}{2} \tag{2.23}$$

From Eqs. (2.22) and (2.23),

$$p_U - p_D = \frac{\rho_a \left(V^2 - V'^2\right)}{2} \tag{2.24}$$

Substituting the above expression for $(p_U - p_D)$ in Eq. (2.21),

$$F = \frac{\rho_a A_T \left(V^2 - V'^2\right)}{2} \tag{2.25}$$

Comparing Eqs. (2.20) and (2.25) we get

$$V_T = \frac{\left(V + V'\right)}{2} \tag{2.26}$$

Thus the velocity of the wind stream at the rotor section is the average of the velocities at its upstream and downstream sides.

At this stage, we introduce a parameter, termed as the axial induction factor into our analysis. The axial induction factor a indicates the degree with which the wind velocity at the upstream of the rotor is slowed down by the turbine. Thus

$$a = \frac{V - V_T}{V} \tag{2.27}$$

From Eqs. (2.26) and (2.27),

$$V_T = V(1-a) \tag{2.28}$$

and

$$V' = V(1-2a) \tag{2.29}$$

As we have seen earlier, the power imparted to the wind turbine is due to the transfer of kinetic energy from the air to the rotor. The mass flow through the rotor over a unit time is

$$m = \rho\, A_T\, V_T \tag{2.30}$$

Hence the power developed by the turbine due to this transfer of kinetic energy is

$$P_T = \frac{1}{2}\rho_a\, A_T\, V_T\, (V^2 - V'^2) \tag{2.31}$$

Substituting for V_T and V' from Eqs. (2.28) and (2.29), we get

$$P_T = \frac{1}{2}\rho_a\, A_T\, V^3\ 4a(1-a)^2 \tag{2.32}$$

Comparing Eq. (2.32) with the expression for power coefficient in Eq. (2.8), we can see that

$$C_P = 4a(1-a)^2 \tag{2.33}$$

For C_p to be maximum,

$$\frac{dC_p}{da} = 0 \tag{2.34}$$

Thus differentiating Eq. (2.33), equating it to zero and solving, we get a=1/3.

Substituting for a in Eq. (2.33), the maximum theoretical power coefficient of a horizontal axis wind turbine is 16/27 and the maximum power produced is

$$P_{T\,MAX} = \frac{1}{2}\rho_a\, A_T\, V^3\ \frac{16}{27} \tag{2.35}$$

This limit for the power coefficient is known as the Betz limit.

It should be noted that, several assumptions are involved in this analysis. Some of these may be questionable when we consider the real flow conditions around a wind turbine. For example, the practical rotor has finite number of blades and the aerodynamic drag and tip losses cannot be neglected. Further, the flow ahead and behind the rotor is not completely axial as assumed under the ideal condition. When the fluid applies torque to the rotor, as a reaction, rotational wake is generated behind the rotor as shown in Fig. 2.16. This will cause energy loss and reduce the peak power coefficient.

Considering the tangential flow behind the rotor, we introduce another factor termed as the tangential induction factor a' in the analysis such that:

$$a' = \frac{\omega}{2\,\Omega} \tag{2.36}$$

Here ω is the induced tangential angular velocity of flow and Ω is the angular velocity of the rotor. Consider an annular stream tube of thickness dr at a distance r from the root of the blade as in the figure. Area of this annular element is

$$A = 2\pi\, r\, dr \tag{2.37}$$

Hence, the thrust force experienced by the annular element may be expressed as

$$dF = 4a\,(1-a)\,\frac{1}{2}\rho_a V^2\; 2\pi\, rdr \tag{2.38}$$

Similarly, the torque on the annular element can be given by

$$dT = 4a'\,(1-a)\,\frac{1}{2}\rho_a V\;\; 2\pi\, rdr\;\; \Omega r\;\; r \tag{2.39}$$

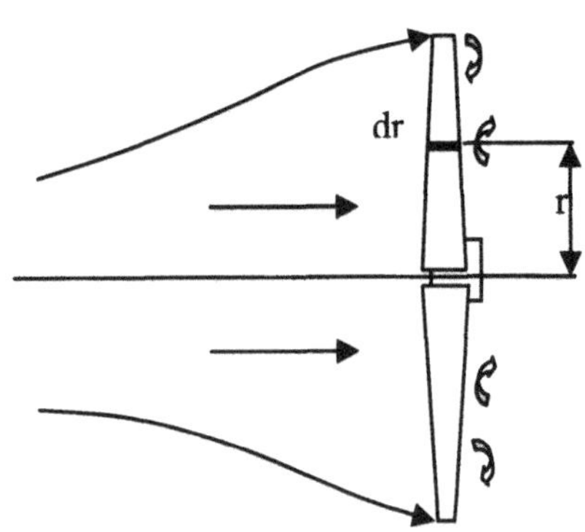

Fig. 2.16. Rotational wake behind the rotor

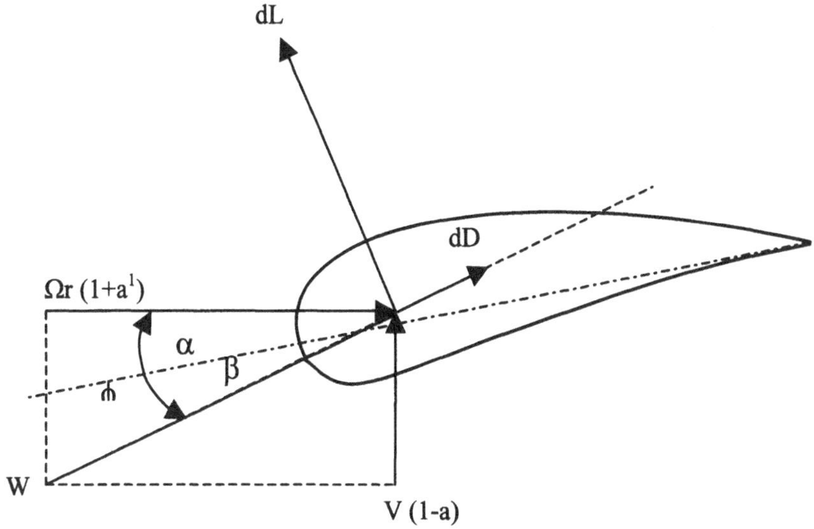

Fig. 2.17. An infinitesimal element of the rotor blade

The power developed by the rotor is the product of this annulus torque and angular velocity, integrated over the total blade span. Thus the power is given by

$$P = \int_0^R \Omega\, 4a' \,(1-a)\,\frac{1}{2}\rho_a V\; 2\pi r dr\; \Omega r\, r \tag{2.40}$$

and the power coefficient is given by

$$C_P = \frac{2}{\rho_a\, A_T\, V^3}\int_0^R \Omega\;\, 4a'\,(1-a)\;\,\frac{1}{2}\rho_a V\;\, 2\pi r dr\;\, \Omega r\, r \tag{2.41}$$

Blade element theory

Blade element theory was initially proposed by Froude and Taylor. In this approach it is considered that the blades are made up of a number of strips arranged in the span wise direction. The strips have infinitesimal thickness. These strips are aerodynamically independent and do not have any interference between them. Under this analysis, the lift and drag forces acting over the strip are estimated and integrated over the total blade span incorporating the velocity terms, to obtain the torque and power developed by the blade. This is further multiplied by the number of blades to get the total rotor torque and power. The blade element theory gives us more understanding on the relationship between the airfoil properties, thrust experienced by the rotor and the power produced by it.

The velocities and forces acting on an infinitesimal blade element are shown in Fig. 2.17. The undisturbed wind velocity V is slowed down to (1-a)V as it reaches the rotor. A velocity of Ω r (1+a$'$) is experienced by the element due to the rotation of the blades and the wake behind the rotor, where a$'$ is the axial induction factor. W represents the resultant of these two velocities. α is the angle of attack, ϕ is the flow angle and β is the blade setting angle.

Lift forces acting on the infinitesimal airfoil section is given by:

$$dL = \frac{1}{2}\rho_a \, C \, dr \, W^2 \, C_L \tag{2.42}$$

where C is the airfoil chord, dr is the thickness of the section considered and C_L is the lift coefficient. Similarly, the elemental drag is given by

$$dD = \frac{1}{2}\rho_a \, C \, dr \, W^2 \, C_D \tag{2.43}$$

where C_D is the drag coefficient.

We can see that the thrust (dF) and torque (dT) acting on the blade element are

$$dF = dL \cos\phi + dD \sin\phi \tag{2.44}$$

and

$$dT = r\left[dL \sin\phi - dD \cos\phi\right] \tag{2.45}$$

From Eqs. (2.42) to (2.45), and considering the total number of blades B,

$$dF = \frac{1}{2}\rho_a \, B \, C \, dr \, W^2 \left[dL \cos\phi + dD \sin\phi\right] \tag{2.46}$$

and

$$dT = \frac{1}{2}\rho_a \, B \, C \, r \, dr \, W^2 \left[C_L \sin\phi - C_D \cos\phi\right] \tag{2.47}$$

From the geometry of the figure we see that,

$$\sin\phi = \frac{V(1-a)}{W} \tag{2.48}$$

and

$$\cos\phi = \frac{\Omega \, r\left(1+a'\right)}{W} \tag{2.49}$$

Similarly, W can be represented as the resultant of the velocities such that

$$W = \left[V^2 (1-a)^2 + \Omega^2 r^2 \left(1+a'\right)^2\right]^{\frac{1}{2}} \tag{2.50}$$

The total torque developed by the rotor can be computed by integrating the elemental torque from the root to the tip of the blade. Rotor power is estimated by multiplying this torque by its angular velocity. The blade element theory can further be combined with the axial momentum theory for further understanding of rotor behavior.

Strip theory

The strip theory is the combination of the momentum theory and the blade element theory. The strip theory can give us more insight to the design parameters of a wind turbine. We have seen that, according to the momentum theory, the axial force on the annulus ring is

$$dF = 4a\left(1-a\right)\frac{1}{2}\rho_a V^2 \, 2\pi r \, dr \tag{2.51}$$

As per the blade element theory,

$$dF = \frac{1}{2}\rho_a \; B \, C \; dr \, W^2 \left[dL \, \cos\phi + dD \, \sin\phi\right] \tag{2.52}$$

From the geometry of Fig. 2.17, the resultant velocity W can be expressed as

$$W = \frac{(1-a)\, V}{\sin\phi} \tag{2.53}$$

Combining Eqs. (2.51), (2.52) and (2.53) and rearranging, we get

$$\frac{a}{(1-a)} = \frac{B\,C}{8\,\pi\,r}\,\frac{\left[dL\,\cos\phi + dD\,\sin\phi\right]}{\sin^2\phi} \tag{2.54}$$

Defining the local solidity σ_r as

$$\sigma_r = \frac{B\,C}{2\,\pi\,r} \tag{2.55}$$

we get that

$$\frac{a}{(1-a)} = \frac{\sigma_r}{4}\,\frac{\left[dL\,\cos\phi + dD\,\sin\phi\right]}{\sin^2\phi} \tag{2.56}$$

Now let us consider the expressions for the elemental torque according to the two theories. Momentum theory gives us the torque dT as

$$dT = 4a'\,(1-a)\;\frac{1}{2}\rho_a V\;2\pi r dr\;\; r\Omega\, r \tag{2.57}$$

whereas, as per the blade element theory, dT can be expressed as

$$dT = \frac{1}{2}\,\rho_a\; B\; C\; r\, dr\; W^2\left[C_L\;\sin\phi - C_D\;\cos\phi\right] \tag{2.58}$$

From the geometry of Fig. 2.17, W can be expressed as

$$W = \frac{(1-a)\;V}{\sin\phi} = \frac{(1+a')\;\Omega\;r}{\cos\phi} \tag{2.59}$$

Thus

$$W^2 = \frac{(1-a)\;(1+a')V\,\Omega\;r}{\sin\phi\;\cos\phi} \tag{2.60}$$

Combining Eqs. (2.57), (2.58) and (2.60) and simplifying,

$$\frac{a'}{(1+a')} = \frac{\sigma_r}{4}\;\frac{\left[dL\;\sin\phi + dD\;\cos\phi\right]}{\sin\phi\;\cos\phi} \tag{2.61}$$

Thus we arrive at an expression for the tangential induction factor. The power developed can be given by

$$P = \int_0^R \Omega\;dT \tag{2.62}$$

and the power coefficient can be computed as

$$C_P = \frac{2}{\rho_a\;A_T\;V^3}\int_0^R \Omega\;dT \tag{2.63}$$

The theories discussed above deals with the basic aerodynamics of wind turbines. Several assumptions are made in our analysis to represent ideal flow around the rotor. These theories are further modified incorporating correction factors to reflect practical flow conditions. Several other models are also available for defining the aerodynamic properties of wind turbines. Examples are approaches based on vortex and cascade theories. Interested readers may go through [3], [4], [5], [9] and [13] for more rigorous analysis on rotor aerodynamics.

2.6 Rotor design

Designing a wind energy conversion system is a complex process. The environmental conditions to which the turbine is exposed can be severe and unpredictable. Principles of aerodynamics, structural dynamics, material science and economics are to be applied to develop machines which are reliable, efficient and cost effective. Several computational models are available for handling this complex design process. Some examples are PROP [9, 12, 6] and WTPERF [2]. In this section, a simple procedure for an approximate design of a wind rotor is discussed, based on the fundamental aerodynamic theories. Following input parameters are to be identified for such a design.

1. Radius of the rotor (R)
2. Number of blades (B)
3. Tip speed ratio of the rotor at the design point (λ_D)
4. Design lift coefficient of the airfoil (C_{LD})
5. Angle of attack of the airfoil lift (α)

The radius of the rotor primarily depends on the power expected from the turbine and the strength of the wind regime in which it operates. Various losses involved in the energy conversion process are also to be considered. If the power expected from the turbine (P_D) at its design point is known, with the relationship

$$P_D = \frac{1}{2} C_{PD}\, \eta_d \eta_g\, \rho_a\, A_T\, V_D^3 \tag{2.64}$$

the radius of the rotor can be estimated as

$$R = \left[\frac{2\, P_D}{C_{PD}\, \eta_d \eta_g\, \rho_a\, \pi\, V_D^3} \right]^{\frac{1}{2}} \tag{2.65}$$

where C_{PD} is the design power coefficient of the rotor, η_d is the drive train efficiency, η_g is the generator efficiency and V_D is the design wind velocity. For a well designed system, the design power coefficient (C_{PD}) may be in the range of 0.4 to 0.45 and the combined efficiency of the drive train and generator may be taken as 0.9. If the design is to be based on the energy required for a specific application (E_A), the rotor radius can be calculated by

$$R = \left[\frac{2\, E_A}{\eta_S\, \rho_a\, \pi\, V_M^3\, T} \right]^{\frac{1}{2}} \tag{2.66}$$

where η_S is the overall system efficiency, V_M is the mean wind velocity over a period and T is the number of hours in that period. For example, if our design is based on the daily energy demand, then V_M is the daily mean wind velocity and T= 24. For applications like water pumping, it is reasonable to take η_S in the range of 0.12 to 0.15. For wind electric generation, η_S may vary from 0.25 to 0.35, depending on the design features of the system.

Design tip speed ratio depends on the application for which the turbine is being developed. For example, when we design the rotor for a wind pump which require high starting torque, low tip speed ratio is chosen. On the other hand, if our

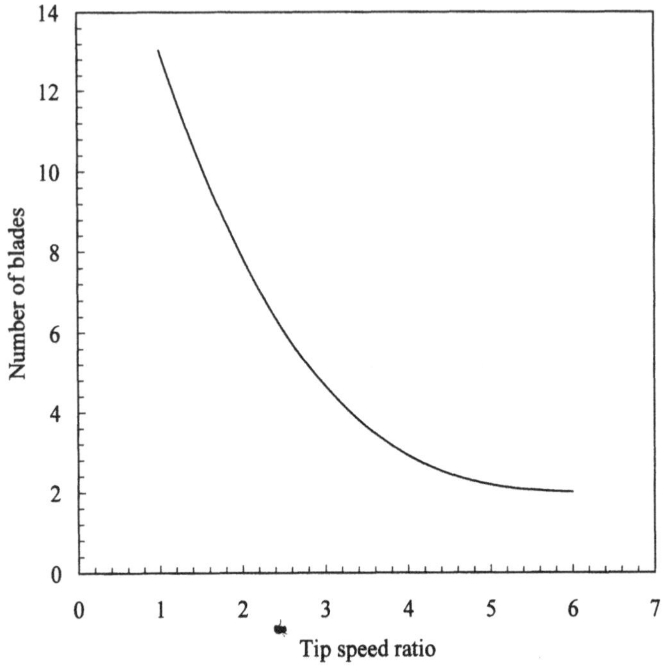

Fig. 2.18. Number of blades and design tip speed ratio

intention is to generate electricity, we require a fast running rotor and hence high tip speed ratio. Most of the water pumping wind mills coupled with piston pumps has λ_D ranging from 1 to 2. For wind pumps with roto-dynamic pumps, we may go for higher λ_D. For aero generators λ_D may be higher than 5.

Number of blades in a rotor is directly related to the design tip speed ratio. The higher the tip speed ratio, the lower would be the number of blades. Fig. 2.18 gives a guideline for choosing the number of blades based on the design tip speed ratio.

Angle of attack α and corresponding lift coefficient C_{LD} are taken from the performance data of the airfoil used in the design. Data on drag and lift coefficients at different angles of attack may be available for standard airfoil sections. As we would like to minimize the drag force and maximize the lift, α corresponding to minimum C_D/C_L is to be identified. This can be obtained from a C_D-C_L curve of an airfoil by plotting a tangent to the curve from the origin as shown in Fig. 2.19. If such data are not readily available for the selected airfoil, the required information should be generated through wind tunnel experiments. Once all these input parameters are identified, the design procedure involves computing the chord length C and blade setting angle β at different sections of the airfoil. C and β of a blade section at a distance r from the center can be determined by the following sets of relationships [6]:

$$\lambda_r = \frac{\lambda_D\, r}{R} \tag{2.67}$$

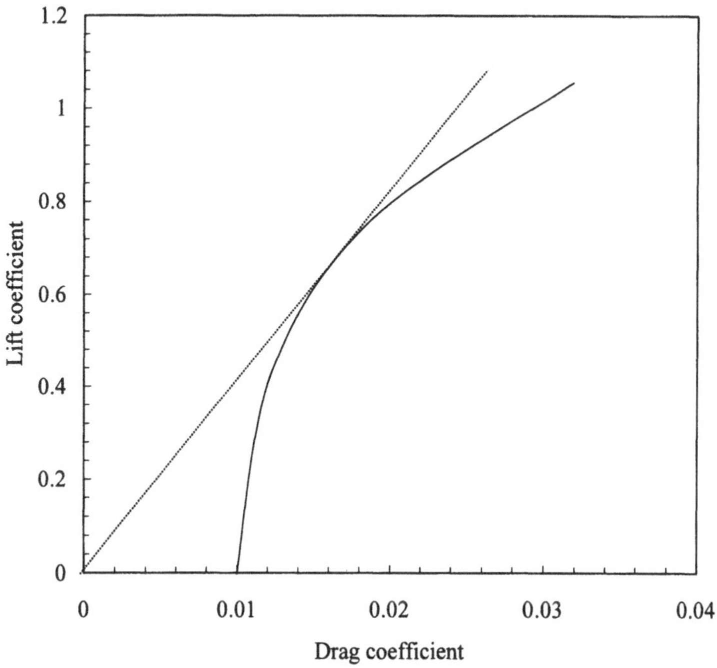

Fig. 2.19. C_D-C_L relationship of an airfoil

$$\phi = \frac{2}{3}\tan^{-1}\frac{1}{\lambda_r} \tag{2.68}$$

$$\beta = \phi - \alpha \tag{2.69}$$

$$C = \frac{8\pi r}{B C_{LD}}\left(1 - \cos\phi\right) \tag{2.70}$$

Example

Design the rotor for an aero generator to develop 100 W at a wind speed of 7 m/s. NACA 4412 airfoil may be used for the rotor.

Let us take the design power coefficient as 0.4 and the combined drive train and generator efficiency 0.9. Taking the air density as 1.224 kg/m^3, from Eq. (2.65), the rotor radius is

$$R = \left[\frac{2\times 100}{0.4\times 0.9\times 1.224\times \pi\times 7^3}\right]^{\frac{1}{2}} = 0.65\,m$$

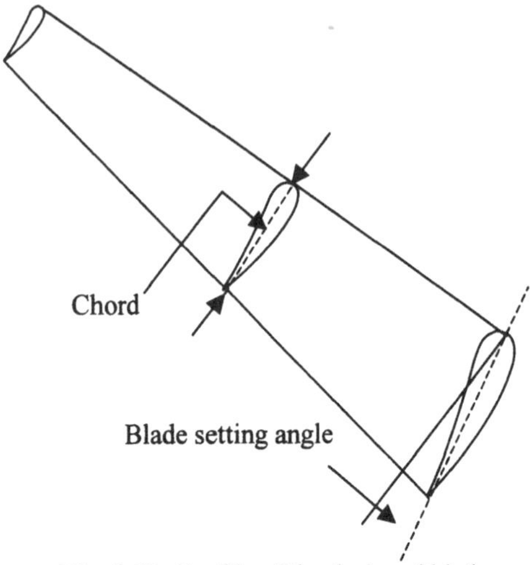

Fig. 2.20. Profile of the designed blade

As the application is to generate electricity, we prefer a low solidity rotor with minimum number of blades and working at high tip speed ratio. For aerodynamic and structural stability, a three bladed rotor is considered. From Fig. 2.18, it is logical to take a design tip speed ratio of 5 for such rotors. From the available performance data of NACA 4412 airfoil, it is seen that minimum C_D/C_L of 0.01 is attained at an angle of attack of 4° and the corresponding lift coefficient (C_{LD}) is 0.8.

The total blade length, starting from 0.2 R to R, is divided in to 9 sections at an interval of 6.5 cm. The chord and blade setting angle at these sections are calculated using Eqs.(2.67) to (2.70). The results are given in Table 2.1. Profile of the designed blade is shown in Fig. 2.20.

Table 2.1. Design specifications of the wind rotor

Section	Sectional radius (m)	Angle of attack (degrees)	Design Lift coefficient	Chord (m)	Blade Setting Angle (degrees)
1	0.13	4	0.8	0.182	26.015
2	0.195	4	0.8	0.155	18.471
3	0.26	4	0.8	0.129	13.719
4	0.325	4	0.8	0.109	10.542
5	0.39	4	0.8	0.094	8.296
6	0.455	4	0.8	0.082	6.636
7	0.52	4	0.8	0.072	5.362
8	0.585	4	0.8	0.065	4.357
9	0.65	4	0.8	0.059	3.544

It can be observed that, to keep a constant angle of attack and thus the same lift throughout the blade, we had to vary the chord and blade setting angle throughout the blade length. Blade material may be wasted in such a design. Fabrication of such a profile may also be difficult. These problems can be avoided by keeping constant chord throughout the blade length and linearizing the blade profile.

Example

A wind pump has to develop 1.75 kWh per day for lifting water for irrigation. Daily mean wind speed at the site is 4 m/s. Design the rotor for the wind pump. 1750 Wh energy is to be developed in a day. Taking air density as 1.224 kg/m^3 and overall system efficiency as 0.15, from Eq. (2.66), the rotor radius is:

$$R = \left[\frac{2\times 1800}{0.15\times 1.224\times \pi\times 4^3\times 24}\right]^{\frac{1}{2}} \cong 2\,m$$

Let us make the rotor with curved steel plates. Such blades with 10 per cent curvature show reasonably good aerodynamic properties at low Reynolds numbers. The chord may be fixed as 0.30 m. As we are using this rotor for water pumping, the rotor must show high starting torque. Hence we should go for a high solidity rotor with more number of blades. Let us design the rotor with 6 blades. From Fig. 2.18, it is reasonable to take a design tip speed ratio of 2 for such systems.

In the previous design with varying chord, we could keep the angle of attack and lift coefficient constant throughout the blade. In the present case, as the chord is fixed, the angle of attack and thus the lift coefficient at different blade sections would vary. From Eq. (2.70), the lift coefficient at various sections can be calculated by

$$C_L = \frac{(8\pi r)}{BC}\left\{1-\cos\left(\frac{2}{3}\tan^{-1}\left(\frac{R}{\lambda_D\, r}\right)\right)\right\} \tag{2.71}$$

To find the angle of attack corresponding to these lift coefficients, we should have the α-C_L relationship of the selected blade. Once the angles of attack at various

Table 2.2 Specifications of the wind pump rotor

Section No	R (m)	λ_r	C (m)	ϕ	C_L	α	β
1	0.5	0.625	0.325	38.68269	1.411659	9.9	28.78269
2	1	1.25	0.325	25.78628	1.281478	6.8	18.98628
3	1.25	1.5625	0.325	21.75719	1.145949	3.6	18.15719
4	1.5	1.875	0.325	18.72448	1.021681	1.9	16.82448
5	1.75	2.1875	0.325	16.38642	0.914779	0.2	16.18642

sections are determined, we can find out the corresponding flow angles and blade setting angles using Eqs. (2.68) and (2.69).

As we have constant chord, let us divide the 2 m blade length into five equal sections. Specification of the rotor designed as per the procedure described above is given in Table 2.2. It should be noted that the calculated blade setting angle does not vary linearly. While fabrication, the twist may further be linearized between the two ends of the blade.

2.7 Rotor performance

Once a wind turbine rotor is designed for a specific application, its performance characteristics are to be brought out before we fabricate the prototype. Dimensionally similar scaled down models of the proposed design are tested under wind tunnels for this purpose.

Low speed wind tunnels working at a very low Mach number (for example 0.3) are used for these experiments. Details of such a tunnel used for testing a wind turbine are shown in Fig. 2.21. The actual field conditions under which a wind turbine operates can be simulated inside the tunnel. Model wind turbine to be tested is fixed at the throat or working section. There are provisions to vary wind velocity as well as the load on the rotor shaft. A complete set of instrumentation with transducers and data loggers sense and record information like wind velocity, rotor torque and speed.

The rotor is tested for different wind velocities and shaft loads. Power-speed and torque-speed characteristics of the rotor, at different wind velocities, can be generated from the test results. This is further used to estimate the CP-λ relationship of the rotor. Starting behavior of the rotor can also be found out.

Example

A wind turbine model of 1 m diameter was tested in a wind tunnel. Test results are given in Table 2.3. Generate the CP-λ curve of the rotor.

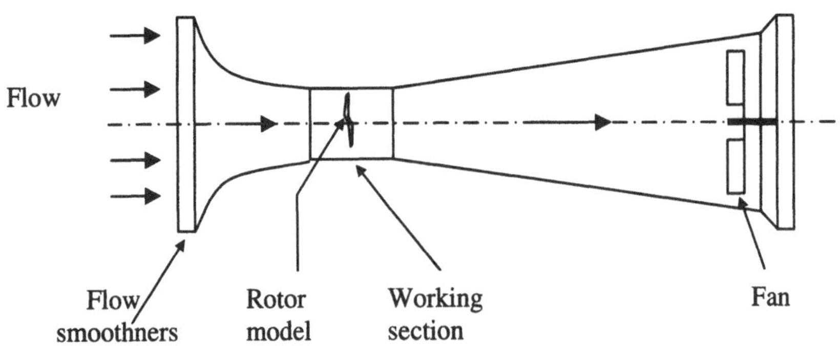

Fig. 2.21. A low speed wind tunnel

Table 2.3. Observations of the wind tunnel test

4 m/s		6 m/s		8 m/s	
Rotor speed (r/min)	Power (W)	Rotor speed (r/min)	Power (W)	Rotor speed (r/min)	Power (W)
306	12.30	482	42.55	673	104.54
352	13.37	550	45.66	764	109.46
397	13.68	619	47.73	856	113.15
443	14.33	688	48.46	948	110.69
489	13.53	757	44.62	1039	100.85
535	11.99	768	43.58	1055	98.39

The power-speed curve of the rotor at 4 m/s, 6 m/s and 8 m/s are shown in Fig. 2.22. The power coefficient of the model rotor can be estimated from the relationship

$$C_P = \frac{8\,P}{\pi\,\rho_a\,d^2\,V^3} \tag{2.72}$$

The tip speed ratio may be calculated as

$$\lambda = \frac{\pi\,D\,N}{V} \tag{2.73}$$

λ and the corresponding C_P for the rotor, at different wind velocities and loads, are calculated using the above expressions and plotted in Fig. 2.23.

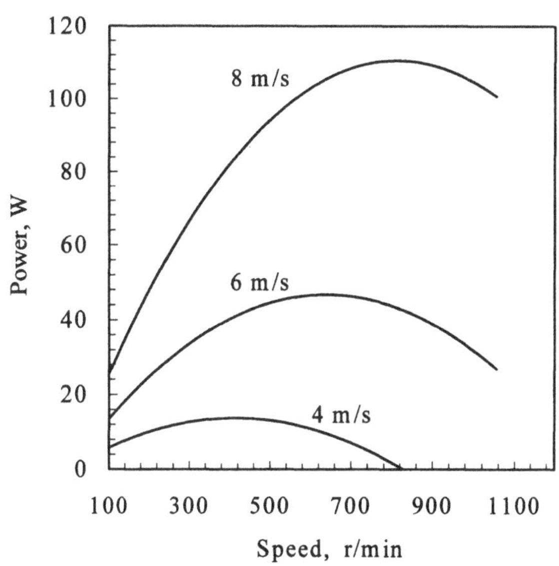

Fig. 2.22. Power- speed curves of the rotor

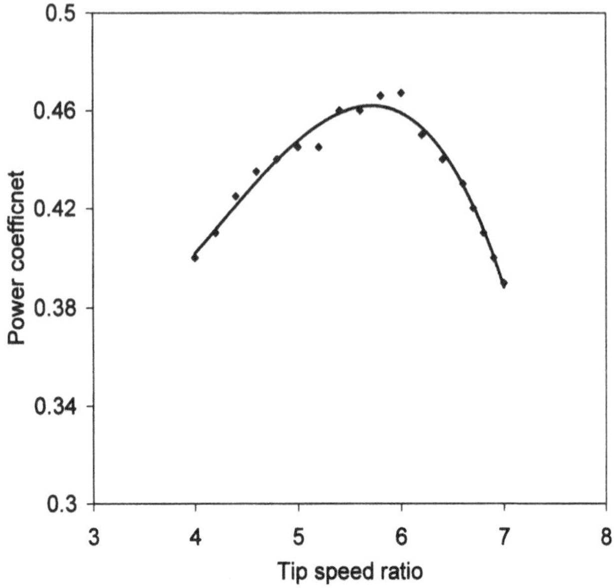

Fig. 2.23. C_P - λ curve of the rotor

The relationship between C_P and λ at all the wind velocities converges to a single curve. This non-dimensional relationship can represent the performance of any dimensionally similar rotor irrespective its size.

References

1. Buhl ML (2000) WT- PERF user's guide. NREL
2. Dahl KS (1999) Experimental verifications of the new RISO–A1 airfoil family for wind turbines. In: Proceedings of EWEC'99, pp 85-88
3. Dini P, Coiro DP, Bertolucci S (1995) Vortex model for airfoil stall prediction using an interactive boundary layer method. In: Wind Energy 1995, SED, American Society of Mechanical Engineers
4. Eggleston DM, Starcher KL (1990) A comparative study of the aerodynamics of several wind turbines using flow visualization. J. Solar Energy Engineering 112: 301-309
5. Hansen AC, Butterfield CP (1993) Aerodynamics of horizontal -axis wind turbines. Annual Review of Fluid Mechanics 25: 115-149
6. Koleffler RG, Sitz EL (1946) Electrical energy from winds. (Kansas State College of Engineering, Experiment Station Bulletin 52, Manhattan, Kansas).
7. Lysen EH (1982) Introduction to wind energy. CWD, Netherlands
8. Mathew S (1990) Design, fabrication and testing of a Savonius rotor with deflector augmentor. M.Tech. thesis, Kerala Agricultural University
9. McCarty J (1993) PROP93 User's Guide. Alternative Energy institute

10. Rohrbach C, Wainauski H, Worabel R (1977) Experimental and Analytical Research on the Aerodynamics of Wind –Driven Turbines. COO-2615-T2, Windsor Locks, CT: Hamiltion Standard
11. Savonius SJ (1931) The S-rotor and its applications. Mechanical Engineering 53 (5):333 - 338.
12. Selig MS, Tangler JL (1995) Development and application of a multipoint inverse design method for horizontal axis wind turbines. Wind Engineering 19 (2): pp 91-105
13. Somers DM, Tangler JL (1996) Wind tunnel test of the S814 thick root airfoil. Trans. of ASME, J. of Solar Energy Engineering 118: 217-221
14. Timmer WA, vanRooy RPJOM (1993) Wind tunnel results for 25% thick wind turbine blade airfoil. Proceedings of EWEC'93 pp 416-419.
15. Wilson RE (1980) Wind turbine aerodynamics. J of Industrial Aerodynamics.5: 358-372
16. Wilson RE, Walker SN (1976) Performance analysis program for propeller type wind turbines. Oregon State University

3 Analysis of wind regimes

The most critical factor influencing the power developed by a wind energy conversion system is the wind velocity. Due to the cubic relationship between velocity and power, even a small variation in the wind speed may result in significant change in power. For example, an increase in the wind speed by 10 per cent may enhance the productivity of the turbine by over 33 per cent. As the wind velocity and thus the power vary from place to place, the first step in planning a wind energy project is to identify a suitable site, having strong and impressive wind spectra.

Wind is stochastic in nature. Speed and direction of wind at a location vary randomly with time. Apart from the daily and seasonal variations, the wind pattern may change from year to year, even to the extent of 10 to 30 per cent. Hence, the behavior of the wind at a prospective site should be properly analyzed and understood. Realizing the nature of wind is important for a designer, as he can design his turbine and its components in tune with the wind characteristics expected at the site. Similarly, a developer can assess the energy that could be generated from his project, if the wind regime characteristics are known.

Average wind velocity gives us a preliminary indication on a site's wind energy potential. A location having good average wind speed - say for example with a minimum of 7 m/s - is expected to be suitable for wind electric generation. However, for a detailed planning, apart from the average strength of the wind spectra, its distribution is also important. Statistical models are being successfully used for defining the distribution of wind velocity in a regime, over a given period of time.

Once the wind velocity and its distribution at a prospective site are available, we can proceed further with the assessment of the energy potential. One of the initial questions to be addressed is "How much energy is available per unit area of the rotor?" Percentage of time for which the wind is within a useful velocity range, the most frequent wind velocity, and the velocity contributing maximum energy to the regime are some other factors of interest. Similarly, for the safe structural design, possibilities of extremely high wind at the site should be identified.

In this chapter, let us discuss the basic nature of wind along with the methods of measuring its strength. Statistical models commonly used for wind resource analysis are described, indicating their application in wind energy conversion. The discussions are further extended to the derivation of methods and indices for assessing the energy potential of a given wind regime.

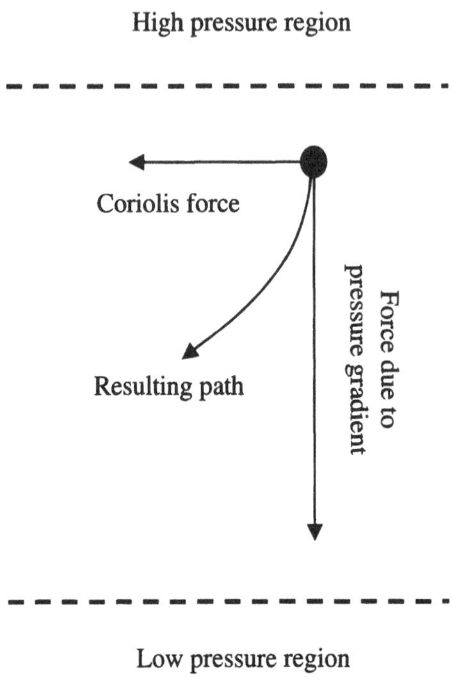

Fig. 3.1. Wind direction affected by the Coriolis force

3.1 The wind

The earth receives around 1.7×10^{14} kW of power from the sun in the form of solar radiation. This radiation heats up the atmospheric air. The intensity of this heating will be more at the equator (0° latitude) as the sun is directly overhead. Air around the poles gets less warm, as the angle at which the radiation reaches the surface is more acute. The density of air decreases with increase in temperature. Thus, lighter air from the equator rises up into the atmosphere to a certain altitude and then spreads around. This causes a pressure drop around this region, which attracts the cooler air from the poles to the equator. This movement of air causes the wind.

Thus, the wind is generated due to the pressure gradient resulting from the uneven heating of earth's surface by the sun. As the very driving force causing this movement is derived from the sun, wind energy is basically an indirect form of solar energy. One to two per cent of the total solar radiation reaching the earth's surface is converted to wind energy in this way.

The wind described above, which is driven by the temperature difference, is called the geostrophic wind, or more commonly the global wind. Global winds, which are not affected by the earth surface, are found at higher altitudes. The rotation of earth leads to another phenomenon near its surface called the Coriolis effect, named after the famous mathematician Gustave Gaspard Coriolis. Due to the Coriolis effect, the straight movement of air mass from the high pressure region to the low pressure region is diverted as shown in Fig. 3.1. Under the influence of Coriolis forces, the air move almost parallel to the isobars. Thus, in the northern hemisphere, wind tends to rotate clockwise where as in the southern hemisphere the motion is in the anti-clockwise direction.

3.1.1 Local effects

Changes in velocity and direction of wind near the surface, say up to 100 m above the ground, is more important as far as energy conversion is concerned. In this region, the wind pattern is further influenced by several local factors.

Land and sea breezes are examples for the local wind effects. During the day time, land gets heated faster than the sea surface. As a result, the air near the land rises, forming a low pressure region. This attracts cool air to the land from the sea. This is called the sea breeze. During night time, the process gets reversed as cooling is faster on land. Thus wind blows from the land to the sea, which is called the land breeze.

In mountain valleys, the air above the surface gets heated and rises up along the slopes during the day time. This is replaced by the cool air, resulting in the valley winds. During the night, the flow is from the mountain to the valley which is known as the mountain wind. Quite often, this phenomenon may create very strong air currents, developing powerful wind. Wind shear, turbulence and acceleration over the ridges are some other examples for local wind effects.

3.1.2 Wind shear

The flow of air above the ground is retarded by frictional resistance offered by the earth surface (boundary layer effect). This resistance may be caused by the roughness of the ground itself or due to vegetations, buildings and other structures present over the ground. For example, a typical vertical wind profile at a site is shown in Fig. 3.2. Theoretically, the velocity of wind right over the ground surface should be zero. Velocity increases with height upto a certain elevation. In the above example, the velocity increases noticeably upto 20 m, above which the surface influence is rather feeble.

The rate at which the velocity increases with height depends on the roughness of the terrain. Presence of dense vegetations like plantations, forests, and bushes slows down the wind considerably. Level and smooth terrains do not have much effect on the wind speed. The surface roughness of a terrain is usually represented by the roughness class or roughness height. The roughness height of a surface may be close to zero (surface of the sea) or even as high as 2 (town centers). Some typical values are 0.005 for flat and smooth terrains, 0.025-0.1 for open grass lands, 0.2 to 0.3 for row crops, 0.5 to 1 for orchards and shrubs and 1to 2 for forests, town centers etc.

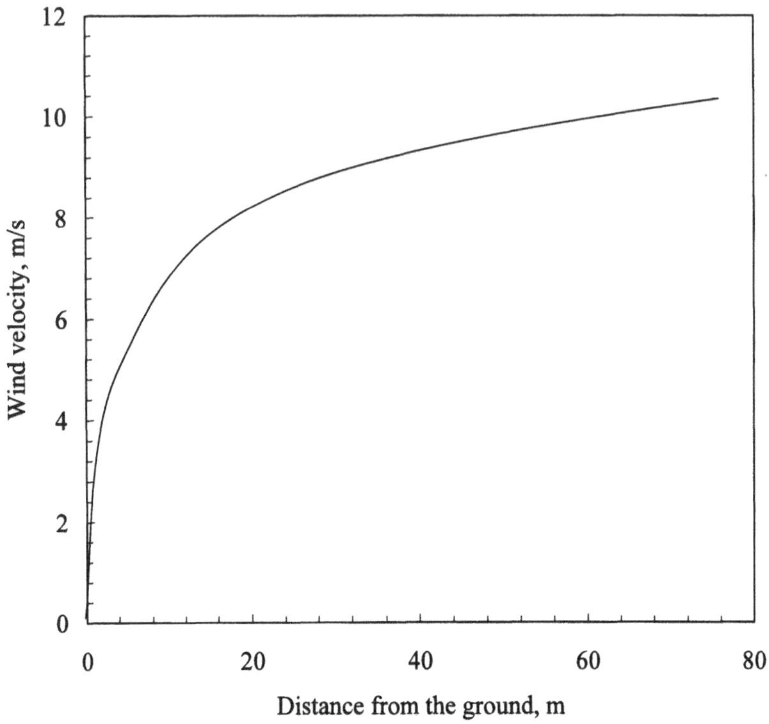

Fig. 3.2. Variation of wind velocity with height

Roughness height is an important factor to be considered in the design of wind energy plants. Suppose we have a wind turbine of 30 m diameter and 40 m tower height, installed over the terrain described in Fig. 3.2. The tip of the blade, in its lower position, would be 25 m above the ground. Similarly, at the extreme upper position, the blade tip is 55 m above the ground. As we see, the wind velocities at these heights are different. Thus, the forces acting on the blades as well as the power available would significantly vary during the rotation of the blades. This effect can be minimized by increasing the tower height.

The wind data available at meteorological stations might have been collected from different sensor heights. In most of the cases, the data are logged at 10 m as per recommendations of the World Meteorological Organization (WMO). In wind energy calculations, we are concerned with the velocity available at the rotor height. The data collected at any heights can be extrapolated to other heights on the basis of the roughness height of the terrain.

Due to the boundary layer effect, wind speed increases with the height in a logarithmic pattern. If the wind data is available at a height Z and the roughness height is Z_0, then the velocity at a height Z_R is given by [11, 13]

$$V(Z_R) = V(Z)\,\frac{\ln\left(Z_R/Z_0\right)}{\ln\left(Z/Z_0\right)} \tag{3.1}$$

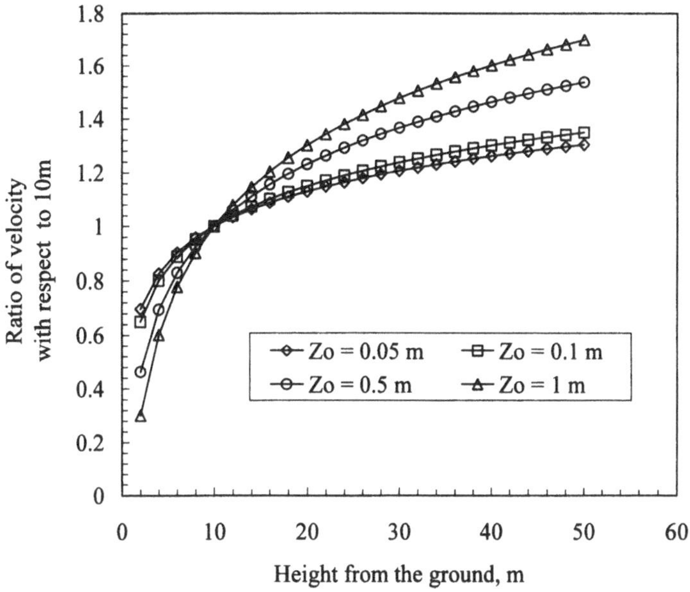

Fig. 3.3. Velocity ratio with respect to 10 m for different roughness heights

where V (Z_R) and V(Z) are the velocities at heights Z_R and Z respectively. Thus, if the velocity of wind measured at a height of 10 m is 7 m/s and the roughness height is 0.1, the velocity at 40 m above the ground is 9.1 m/s. It should be noted that the power available at 40 m is 2.2 times higher than at 10 m. The wind velocities at different heights relative to that at 10 m, as affected by the roughness heights, are shown in Fig. 3.3.

In some cases, we may have data from a reference location (meteorological station for example) at a certain height. This data is to be transformed to a different height at another location with similar wind profile but different roughness height (for example, the wind turbine site). Under such situations, it is logical to assume that the wind velocity is not significantly affected by the surface characteristics beyond a certain height. This height may be taken as 60 m from the ground level [10]. Thus, expressing the velocity at 60 m in terms of the reference location,

$$V(60) = V(Z_R)\left(\frac{\ln\left(60/Z_{OR}\right)}{\ln\left(Z_R/Z_{OR}\right)}\right) \tag{3.2}$$

where Z_{OR} is the roughness height at the reference location. Considering the second location

$$V(60) = V(Z)\left(\frac{\ln\left(60/Z_O\right)}{\ln\left(Z/Z_O\right)}\right) \tag{3.3}$$

Dividing Eq. (3.3) by Eq. (3.2) yields,

$$V(Z) = V(Z_R)\left(\frac{\ln\left(60/Z_{OR}\right)\ \ln\left(Z/Z_O\right)}{\ln\left(60/Z_O\right)\ \ln\left(Z_R/Z_{OR}\right)}\right) \tag{3.4}$$

Example

The wind velocity measured at 10 m height at a meteorological observatory is 7 m/s. Find out the velocity at 40 m height at a wind turbine site having similar wind profile. The roughness heights at the observatory and wind turbine location are 0.03 m and 0.1 m respectively.
Based on Eq. (3.4), we have

$$V(Z) = 7\times\left(\frac{\ln\left(60/0.03\right)\times\ln\left(40/0.1\right)}{\ln\left(60/0.1\right)\times\ln\left(10/0.03\right)}\right) = 8.58\ \text{m/s}$$

3.1.3 Turbulence

The speed and direction of wind change rapidly while it passes through rough surfaces and obstructions like buildings, trees and rocks. This is due to the turbulence generated in the flow. Extent of this turbulence at the upstream and downstream of the flow is shown in Fig. 3.4. The presence of turbulence in the flow not only reduces the power available in the stream, but also imposes fatigue loads on the turbine.

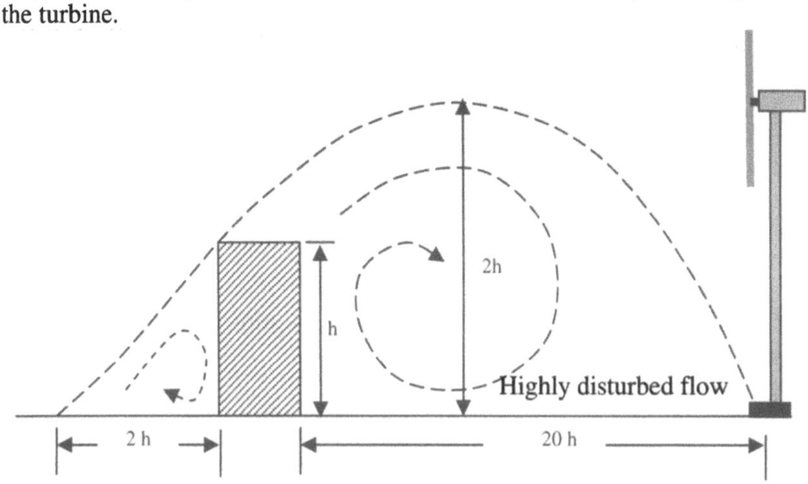

Fig. 3.4. Turbulence created by an obstruction

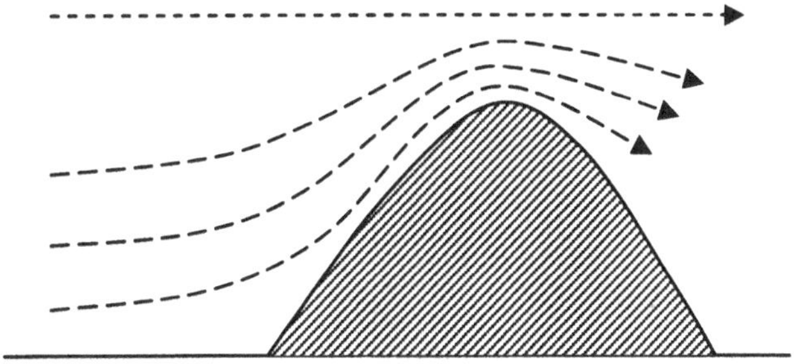

Fig. 3.5. The acceleration effect over ridges

Intensity of the turbulence depends on the size and shape of the obstruction. Based on its nature, the turbulent zone can extend upto 2 times the height of the obstacle in the upwind side and 10 to 20 times in the downwind side. Its influence in the vertical direction may be prominent to 2 to 3 times the obstacle height. Hence before citing the turbine, the obstacles present in the nearby area should be taken in to account. The tower should be tall enough to overcome the influence of the turbulence zone.

3.1.4 Acceleration effect

A smooth ridge, as shown in Fig. 3.5, accelerates the wind stream passing over it. The acceleration is caused by the squeezing of wind layers over the mount as shown in the figure. The degree of acceleration depends on the shape of the ridge. This effect can be fully exploited for energy generation, if the slope of the ridge is between 6° and 16°. Angles greater than 27° and less than 3° are not favorable.

Another important factor is the orientation of the ridge. The acceleration effect is high when the prevailing wind is perpendicular and low when it is parallel to the ridge line. Similarly, if the ridge has a concave side facing the wind, the effect is more desirable. Triangular shaped ridges offer better acceleration followed by the smooth and round geometry. Flat topped ridges may pose the problem of turbulence, especially in the lower region.

Mountain passes are another geographical feature causing acceleration of wind. While the flow passes through the notches in the mountain barriers, due the ventury effect, the wind velocity is enhanced. Geometrical configuration (Width, length, slope etc.) of the pass itself is the major factor determining the degree of this acceleration. A pass between two high hills, oriented parallel to the wind direction, would be a cleverly chosen site for wind turbines. The smoother the surface, the higher will be the acceleration.

3.1.5 Time variation

Velocity and direction of wind change rapidly with time. In tune with these changes, the power and energy available from the wind also vary. The variations may be short time fluctuation, day-night variation or the seasonal variation.

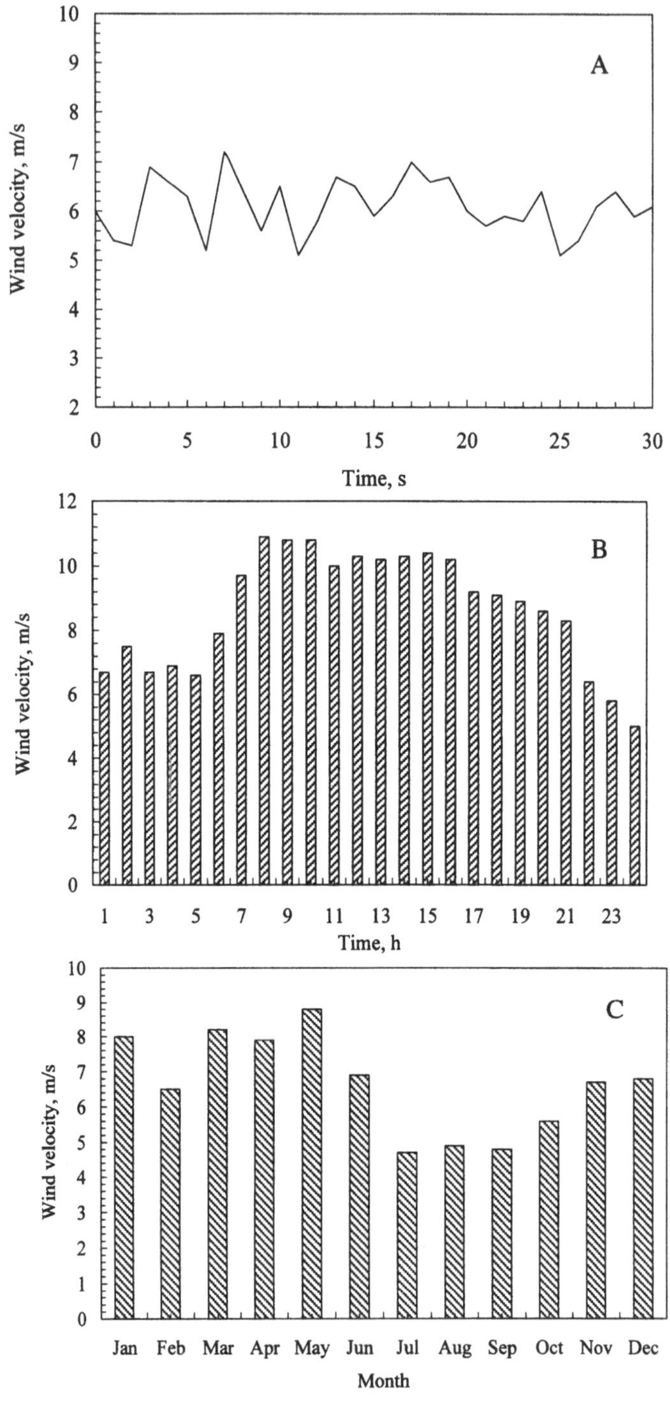

Fig. 3.6. Time variation of wind velocity

An example for the short time variation of wind speed is shown in Fig. 3.6 (A), where the velocity is recorded for 30 s. Here the velocity fluctuates between 5.1 m/s to 7.2 m/s within this time. This short-spanned change in wind speed is primarily due to the local geographic and weather effects. We may experience stronger wind during the day time rather than in night hours. This is termed as the diurnal variation. An example is illustrated in Fig. 3.6 (B). The major reason for the velocity variation here is the difference in temperature between the sea and land surface. It should be noted that the diurnal variation can be advantageous for wind energy generation as we may need more power during the day hours than at night.

Wind speed at a location may also change from season to season as shown in Fig. 3.6 (C). In this case, the period between July to October is more or less lean for wind energy conversion. The root cause for seasonal variation is changes in daylight during the year due to the earth's tilt and elliptical orbit. This effect is more prominent near the poles. Knowledge of these time variations of velocity at a potential wind site is essential to ensure that the availability of power matches with the demand.

3.2 Measurement of wind

A precise knowledge of the wind characteristics at the prospective sites is essential for the successful planning and implementation of wind energy projects. The basic information required for such an analysis is the speed and direction of the prevailing wind at different time scales. Ecological factors may often be helpful in identifying a candidate site for wind power project. Wind data from the nearby meteorological stations can give us a better understanding on the wind spectra available at the site. However, for a precise analysis, the wind velocity and direction at the specific site has to be measured with the help of accurate and reliable instruments.

3.2.1 Ecological indicators

Eolian features may be used as indicators for strength of the wind prevailing in an area. Eolian features are the formations on land surface due to continuous strong wind. Sand dunes are one example for the eolian formation. The sand particles, picked up and carried by the wind flow, are deposited back when the wind speed is lower. The size of the particle thus carried and deposited along with the distance can give us an indication on the average strength of the wind in that region. Other types of eolian features are playa lake, sediment plumes and wind scours. More information on the use of eolian features for locating windy areas is available in [12].

Another way to identify a windy site is to observe the biological indicators. Trees and bushes get deformed due to strong winds. The intensity and nature of this deformation depends on the strength of wind.

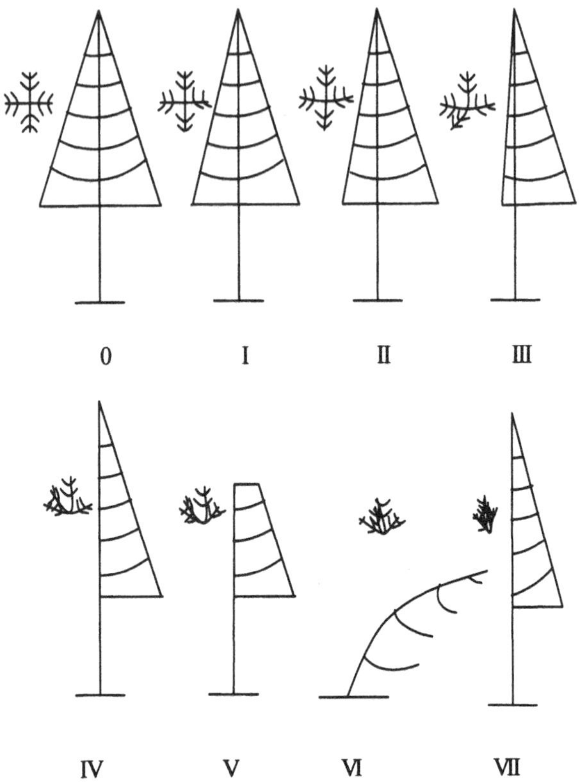

0	No deformation due to wind	IV	Strong flagging
I	Brushing	V	Flagging and clipping
II	Light flagging	VI	Throwing and flagging
III	Moderate flagging	VII	Extreme flagging

Fig. 3.7. Biological rating scales for the wind speed

This method is specifically suitable to judge the wind in valleys, coasts and mountain terrains. Deformations of trees due to wind effect are classified by Putnam [17]. There are five types of deformations under Putnam's classification. They are brushing, flagging, wind throwing, clipping and carpeting.

Wind brushing refers to the leeward bending of branches and twigs of the trees. Brushing may clearly be observed when the trees are off their leaves. This is an indication for light wind, which is not useful for wind energy conversion. In flagging, the branches are stretched out to leeward, with possible stripping of upwind branches. The range of wind speeds corresponding to the flagging effect is of interest for energy conversion. In wind throwing, the main trunk and branches of the tree lean away from the coming wind. This indicates the presence of stronger wind. Under clipping, the lead branches of the tree are suppressed from growing to its normal height due to strong wind. With extreme winds, the trees are clipped even at a very low height. This is termed as wind carpeting.

Based on these deformations, the intensity of wind is rated on a seven point scale put forth by Hewson and Wade [8]. These are illustrated in Fig. 3.7, showing the top and front views of the tree trunk. However, it should be noted that the degree of this deformation may vary from one tree species to the other. For this reason, this method is to be calibrated with long term wind data for a given tree variety. Once such calibration is available, the wind speed range can be directly estimated on the basis of these biological indicators.

3.2.2 Anemometers

The indicators discussed above, along with available wind data from meteorological stations, can give us an idea on the suitability of a given site for wind energy extraction. However, the final selection of the site should be made on the basis of short term field measurements. Anemometers fitted on tall masts are used for such wind measurements. Height of the mast may be the hub height of the turbine to avoid further correction in wind speed due surface shear. As the power is sensitive to the wind speed, good quality anemometers which are sensitive, reliable and properly calibrated should be used for wind measurements.

There are different types of anemometers. Based on the working principle, they can be classified as:

1. Rotational anemometers (cup anemometers and propeller anemometers)
2. Pressure type anemometers (pressure tube anemometers, pressure plate anemometers and sphere anemometers)
3. Thermoelectric anemometers (hot wire anemometers and hot plate anemometers)
4. Phase shift anemometers (ultra sonic anemometers and laser doppler anemometers)

Cup anemometer

The anemometer, most commonly used in wind energy measurements is the cup anemometer. It consists of three (or four) equally spaced cups attached to a centrally rotating vertical axis through spokes (Fig. 3.8). The cups are hemispherical or conical in shape and made with light weight material. This is basically a drag device. When kept in the flow, the wind exerts drag force on the cups. The drag force is given by

$$F_D = C_D \frac{1}{2} A \rho_a V^2 \tag{3.5}$$

where C_D is the drag coefficient, A is the area of cup exposed to the wind, ρ_a is the air density and V is the wind velocity.

As the drag coefficient of concave surface is more than on the convex surface, the cup with its concave side facing the wind experiences more drag force. This causes the cups to rotate on its central axis. The intensity of rotation is directly proportional to the velocity of incoming wind.

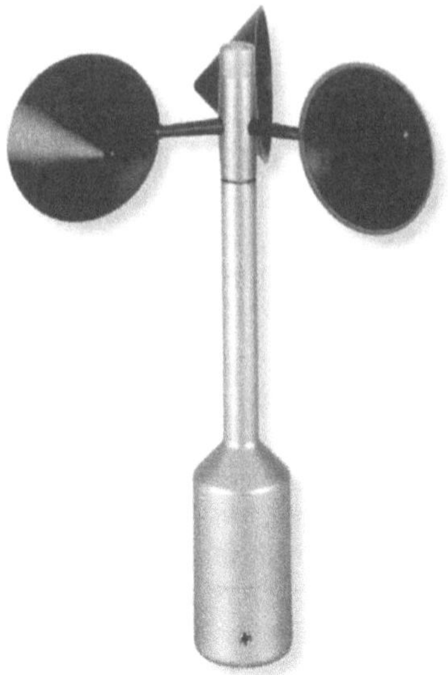

Fig. 3.8. Cup anemometer (Courtesy of THALES instruments GmbH, Werftweg 15, 26135 Oldenburg, Germany, www.thales-instruments.de)

This is further calibrated in terms of wind velocity, which can be directly sensed and recorded.

Although these anemometers can sustain a variety of harsh environments, they have some limitations. It accelerates quickly with the wind but retards slowly as wind ceases. Due to this slow response, cup anemometers do not give reliable measurement in wind gusts. As the drag force is proportional to the density, any changes in the air density will affect the accuracy of metered velocity. Nevertheless of these limitations, cup anemometers are widely used for measuring wind velocity in meteorological as well as wind energy applications.

Propeller anemometer

A propeller type anemometer (aero-vane) consists of a four bladed propeller. The blades are made with light weight materials like aluminum or carbon fiber thermo plastic (CFT). These devices work predominantly on lift force. With an air flow parallel to its axis, the propeller blades experiences a lift force, which turns the propeller at a speed proportional to the wind velocity. For measuring the horizontal and vertical components of wind, three propellers may be fixed on a common mast. The response to any deviation in the direction of wind from the propeller axis follows the cosine law. This means that the velocity perpendicular to the propeller axis would be sensed as zero.

Pressure plate anemometer

The first anemometer of any kind is the pressure plate anemometer. It was invented by Leon Battista Alberti as early as in 1450. This is further refined by

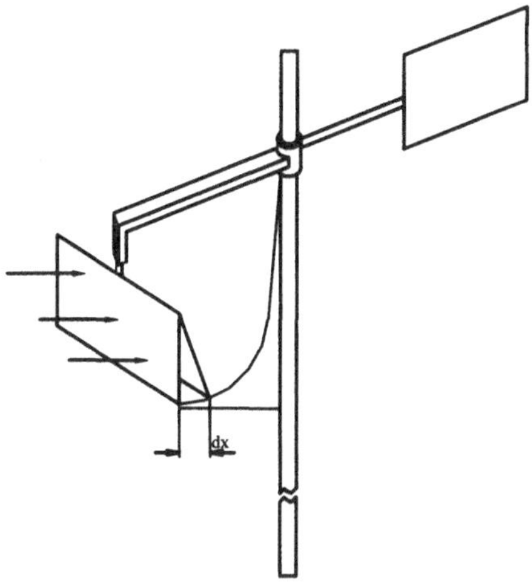

Fig. 3.9. Principle of pressure plate anemometer

Robert Hooke (1664) and Rojer Pickering (1744). It basically consists of a swinging plate held at the end of a horizontal arm. This is attached to a vertical shaft around which the arm can rotate freely (Fig. 3.9). A wind vane directs the plate always perpendicular to the wind flow. As the drag coefficient of a flat plate can be taken as unity, referring Eq. (3.5), the pressure exerted on the plate P by the wind is given by

$$P = \frac{1}{2} \rho_a V^2 \tag{3.6}$$

This pressure makes the plate to swing inward. As the distance through which the plate swings depends on the wind strength, it can be directly calibrated in terms of the wind velocity. Pressure plate anemometers are suited for measuring gusty winds.

Pressure tube anemometers

Another type of anemometer which uses the wind pressure to measure the velocity is the pressure tube anemometer. This works on the principle that, the wind flow

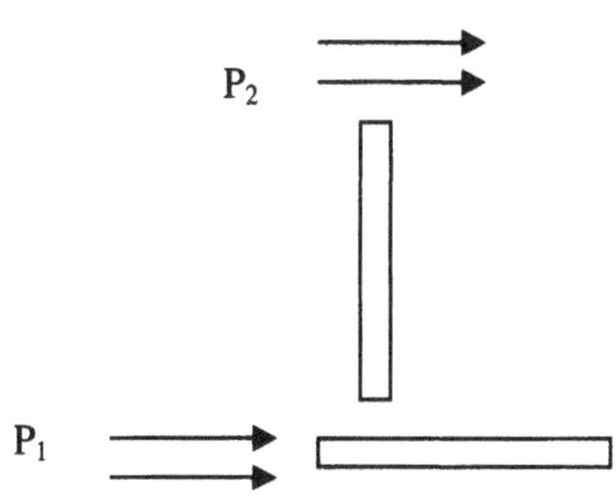

Fig. 3.10. Principle of pressure tube anemometer

passing through the tube creates pressure where as the flow across a tube results in suction. Consider two tubes as shown in the Fig. 3.10. The pressure in the tube parallel to the wind is the sum of atmospheric pressure and the wind pressure. Thus

$$P_1 = P_A + C_1 \frac{1}{2} \rho_a V^2 \tag{3.7}$$

Similarly in the tube perpendicular to the wind, the pressure is

$$P_2 = P_A - C_2 \frac{1}{2} \rho_a V^2 \tag{3.8}$$

where P_A is the atmospheric pressure and C_1 and C_2 are coefficients. Subtracting P_2 from P_1 and solving for V, we get

$$V = \left[\frac{2\,(P_1 - P_2)}{\rho_a\,(C_1 + C_2)} \right]^{0.5} \tag{3.9}$$

Thus, by measuring the difference in pressure inside the two tubes, the wind velocity can be estimated. Values of C_1 and C_2 are available with the instrument. The pressure is measured using standard manometers or pressure transducers. The major advantage of pressure tube anemometer is that it does not have any moving parts. This anemometer has limited application in the open field measurements as the presence of dust, moisture and insects can affect its accuracy.

Sonic anemometer

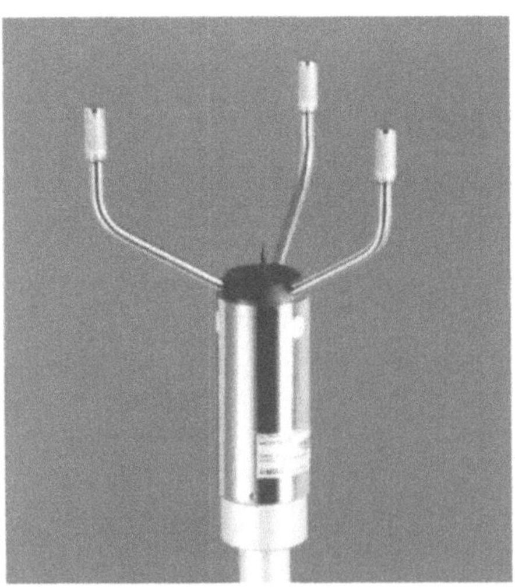

Fig. 3.11. Sonic anemometer (Courtesy of Vaisala Oyj, Vanha Nurmijärventie 21, FIN – 01670, Vantaa, Finland, www.vaisala.com)

Sonic anemometers measure the wind velocity by sensing the changes in the speed of sound in air. It has three arms, mounted perpendicular to each other, as shown in Fig. 3.11. Transducers fitted at the tips of each arm emit acoustic signals which travel up and down through the air. Speed of sound in moving air is different from that through still air. Let V_S be the velocity of sound in still air and V is the wind velocity. If both the sound and wind are moving in the same direction, then the resultant speed of sound waves (V_1) is

$$V_1 = V_s + V \tag{3.10}$$

Similarly, if the propagation of the sound waves is opposite to the wind direction, then the resultant velocity of sound (V_2) is

$$V_2 = V_s - V \tag{3.11}$$

From Eq. (3.10) and Eq. (3.11) we get

$$V = \frac{V_1 - V_2}{2} \tag{3.12}$$

Thus, by measuring the speed of sound waves between the transducer tips during its upward and downward travel, the wind velocity can be estimated. Sonic anemometers also do not have any moving parts. They are reliable and accurate for measuring wind velocity in the range of 0 to 65 m/s. However, they are costlier than the other types of anemometers.

Some other anemometers are bridled anemometers and hot wire anemometers. But they are not common in wind energy measurements. The bridled anemometers are similar to the cup anemometers, but have more number of cups (commonly 32). The rotation of the central axis to which the cups are attached is checked by a spring. As the cups tend to rotate due to wind, tension builds up in the spring. Intensity of this tension is translated into the wind speed. Hot wire anemometers utilize the cooling effect of wind for measuring its velocity. The rate at which a hot wire, maintained at a temperature, is cooled depends on the velocity of air moving over it.

Apart from the sensors discussed above, the complete anemometer assembly includes transducers, data loggers and processing units. For example, the cup and propeller anemometers generally drive a mini generator to convert the rotational speed in to electrical signals. Some designs are fitted with dc generators whereas some recent models employ permanent magnet ac generators. The output voltage of the generator is directly proportional to the wind speed. With an analog to digital (A/D) converter, the analog voltage can be converted to digital form which can further be processed and stored by a suitable data processing system. A typical arrangement powered by a solar panel is shown in Fig. 3.12.

The data processing unit in modern wind measurement systems includes an electronic chip on a computer which receives the instantaneous data and averages it over a specific time period. The standard time interval is 10 minutes. The instantaneous turbulences, which are not of interest in wind energy conversion,

Fig. 3.12. The data acquisition unit (Courtesy of THALES instruments GmbH, Werftweg 15, 26135 Oldenburg, Germany, www.thales-instruments.de)

smoothen out within this 10 minutes interval. At the same time, it can take care of the dynamic changes that are significant for wind energy systems.

For reliable operation, periodic calibration is essential for all anemometers. This is done under ideal conditions against a bench mark anemometer which is considered as the standard. Even with proper calibration, it is observed that some errors may creep in to the measurements. One possible cause for error is the tower shadow. The nearby structures, or even the anemometer tower itself, may shade the instrument resulting in faulty readings. To minimize the risk of tower shadow, cylindrical poles supported by guy wires are preferred than lattice towers. Another problem may be missing data due to functional errors of the recording mechanism.

The quality of data measured using an anemometer will depend on the characteristics of the instrument such as the accuracy, resolution, sensitivity, error, response speed, repeatability and reliability. Accuracy of an anemometer is the degree to which the wind is being measured by the instrument. For example, a typical cup type anemometer can have an accuracy of ± 0.3 m/s. Resolution is the smallest change in wind speed that can be detected by the anemometer and sensitivity is the ratio between the output and input signals. Error is the deviation of indicated velocity from the actual velocity. The response speed tells us how quickly the changes in wind speed can be detected by the anemometer. Repeatability indicates the closeness of readings given by the anemometer while repeated measurements are made under identical conditions. Reliability shows the probability that the anemometer works successfully within its given range of wind speeds. The anemometers used for wind measurements should be periodically checked for these properties.

3.2.3 Wind direction

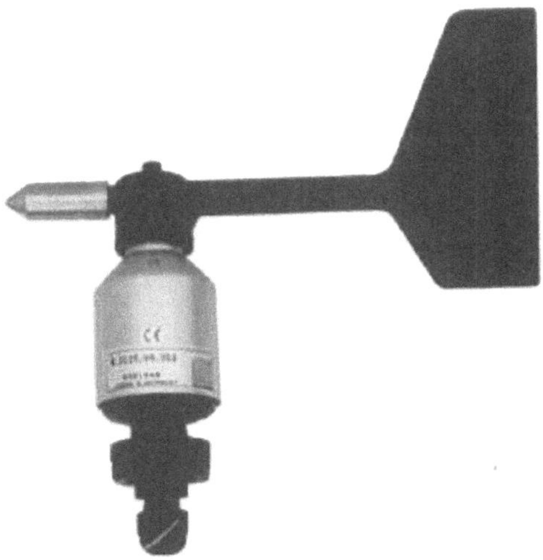

Fig. 3.13. Vanes to measure wind direction (Courtesy of THALES instruments GmbH, Werftweg 15, 26135 Oldenburg, Germany, www.thales-instruments.de)

Direction of wind is an important factor in the siting of a wind energy conversion system. If we receive the major share of energy available in the wind from a certain direction, it is important to avoid any obstructions to the wind flow from this side. Wind vanes were used to identify the wind direction in earlier day's anemometers. However, most of the anemometers used today have provisions to record the direction of wind along with its velocity. A typical arrangement to measure the wind direction used with cup anemometers is shown in Fig. 3.13.

Information on the velocity and direction of wind, in a combined form, can be presented in the wind roses. The wind rose is a chart which indicates the distribution of wind in different directions. The chart is divided into 8, 12 or even 16 equally spaced sectors representing different directions. Three types of information can be presented in a wind rose. (1) The percentage of time for which we receive wind from a particular direction. This can show us the direction from which we get most of our wind. (2) The product of this percentage and the average wind velocity in this direction. This tells us the average strength of the wind spectra. (3) The product of time percentage and cube of the wind velocity. This helps us in identifying the energy available from different directions. Typical wind roses for a location are shown in Fig. 3.14.

3.3 Analysis of wind data

For estimating the wind energy potential of a site, the wind data collected from the location should be properly analyzed and interpreted. Long term wind data from the meteorological stations near to the candidate site

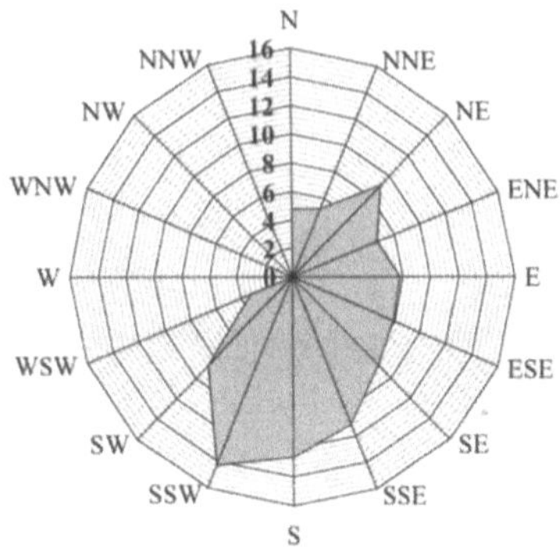

Frequency rose

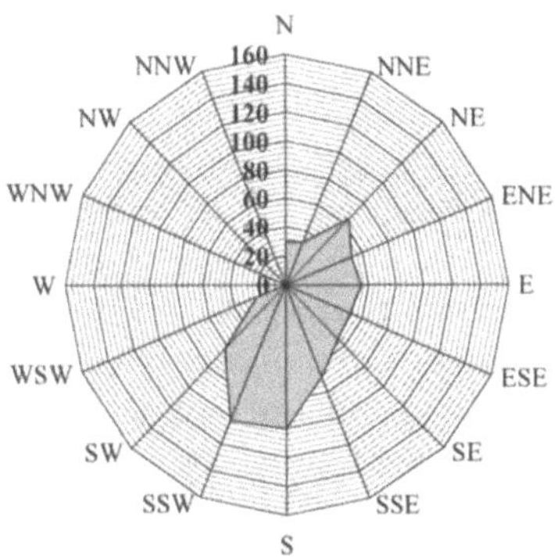

Velocity rose

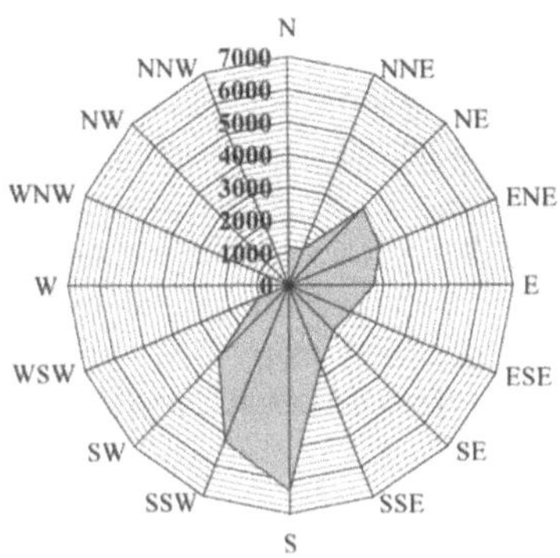

Energy rose

Fig. 3.14. Wind roses showing the distribution of frequency, velocity and energy in different directions

can be used for making preliminary estimates. This data, which may be available for long periods, should be carefully extrapolated to represent the wind profile at the potential site. After this preliminary investigation, field measurements are generally made at the prospective location for shorter periods. One year wind data recorded at the site is sufficient to represent the long term variations in the wind profile within an accuracy level of 10 per cent [6, 16].

Modern wind measurement systems give us the mean wind speed at the site, averaged over a pre-fixed time period. Ten minutes average is very common as most of the standard wind analysis softwares are tuned to handle data over ten minutes. This short term wind data are further grouped and analyzed with the help of models and softwares to make precise estimates on the energy available in the wind. The data are grouped over time spans in which we are interested in. For example, if we want to estimate the energy available at different hours, then the data should be grouped in an hourly basis. The data may also be categorized on daily, monthly or yearly basis.

3.3.1 Average wind speed

One of the most important information on the wind spectra available at a location is its average velocity. In simple terms, the average velocity (V_m) is given by

$$V_m = \frac{1}{n}\sum_{i=1}^{n} V_i \tag{3.13}$$

where V is the wind velocity and n is the number of wind data.

However, for wind power calculations, averaging the velocity using Eq. (3.13) is often misleading. For example, 1 hour wind data from a site collected at 10 minutes interval are shown in Table 3.1. As per Eq. (3.13), the hourly average wind velocity is 6.45 m/s. Taking the air density as 1.24 kg/m^3, the corresponding average power is 166.37 W/m^2. If we calculate the power corresponding to individual velocities and then take the average, the result would be 207 W/m^2. This means that, by calculating the average using Eq. (3.13), we are underestimating the wind power potential by 20 per cent.

Table 3.1. Wind velocity at 10 minutes interval

No	V, m/s	V^3	P, W/m^2
1	4.3	79.51	49.29
2	4.7	103.82	64.37
3	8.3	571.79	354.51
4	6.2	238.33	147.76
5	5.9	205.38	127.33
6	9.3	804.36	498.70

For wind energy calculations, the velocity should be weighed for its power content while computing the average. Thus, the average wind velocity is given by

$$V_m = \left(\frac{1}{n} \sum_{i=1}^{n} V_i^{\,3} \right)^{1/3} \tag{3.14}$$

If we use Eq. (3.14), the average velocity in the previous example is 6.94 m/s and the corresponding power is 207 W/m^2. This shows that due to the cubic velocity-power relationship, the weighted average expressed in Eq. (3.14) should be used in wind energy analysis.

3.3.2 Distribution of wind velocity

Apart from the average strength of wind over a period, its distribution is also a critical factor in wind resource assessment. Wind turbines installed at two sites with the same average wind speed may yield entirely different energy output due to differences in the velocity distribution. For example, consider two sites with a daily wind pattern as shown in Fig. 3.15 (B&C). For the first location, the wind velocity is 15 m/s throughout the day. At the second site, velocity is 30 m/s for the first 12 hours and 0 for the rest of the day. In both the cases, the daily average wind velocity is 15 m/s.

Suppose we install a wind turbine with power curve as shown in Fig. 3.15 (A) at these sites. The turbine will start generating power at its cut-in wind speed of 4 m/s and the generation will be cut-off at 25 m/s. The highest power of 250 kW will be produced at 15 m/s, which is the rated wind speed of the systems (more discussion on the power curve of wind turbines are presented in Chapter-5).

When the turbine is put to work at the first site, as the wind velocity is 15 m/s throughout the day, the system will efficiently work at its rated capacity for all the time, delivering us 6000 kWh. However, in the second case, the turbine will be idle throughout the day as the velocity is 30 m/s in the first half (which is well above the cut-off speed of 25 m/s) and 0 in the second half. The example given here is hypothetical. The real cases would be some where in between the above extreme cases. This shows that, along with the mean wind velocity, the distribution of velocity within the regime is also an important factor in the wind energy analysis.

One measure for the variability of velocities in a given set of wind data is the standard deviation (σ_V). Standard deviation tells us the deviation of individual velocities from the mean value. Thus

$$\sigma_V = \sqrt{\frac{\sum_{i=1}^{n} (V_i - V_m)^2}{n}} \tag{3.15}$$

Lower values of σ_V indicate the uniformity of the data set.

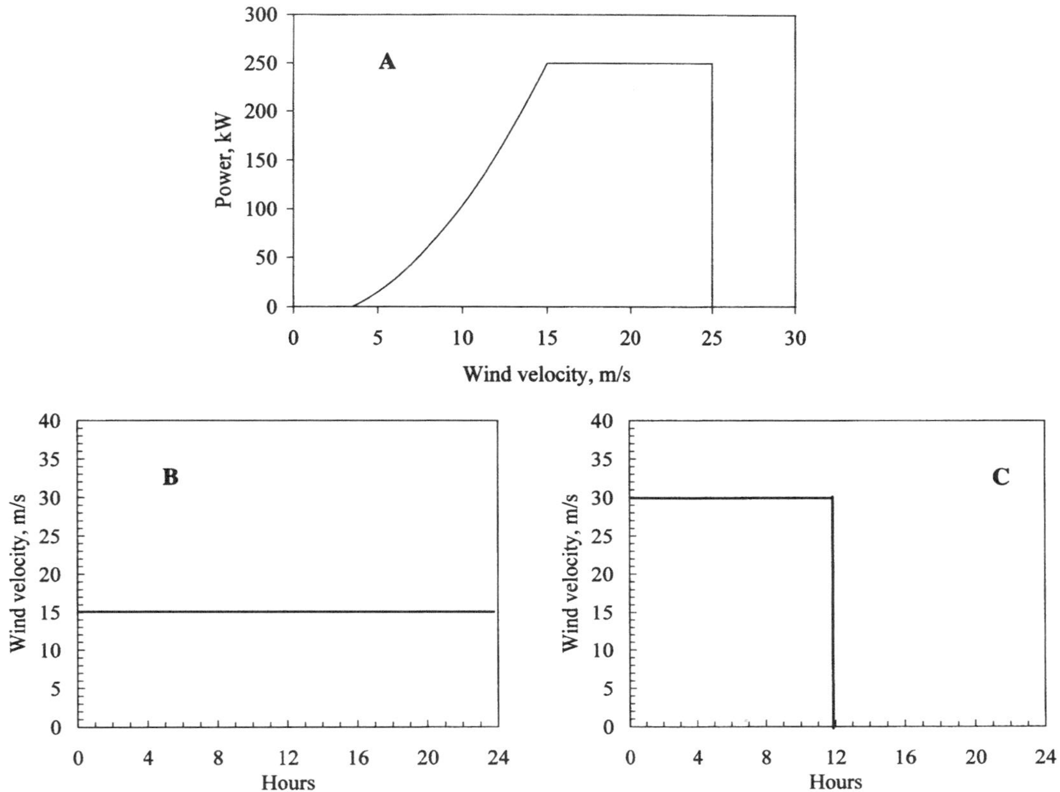

Fig. 3.15. Comparison of wind distribution at two sites

Table 3.2. Frequency distribution of monthly wind velocity

	Velocity, m/s	Hours per month	Cumulative hours
1	0-1	13	13
2	1-2	37	50
3	2-3	50	100
4	3-4	62	162
5	4-5	78	240
6	5-6	87	327
7	6-7	90	417
8	7-8	78	495
9	8-9	65	560
10	9-10	54	614
11	10-11	40	654
12	11-12	30	684
13	12-13	22	706
14	13-14	14	720
15	14-15	9	729
16	15-16	6	735
17	16-17	5	740
18	17-18	4	744

For a better understanding on wind variability, the data are often grouped and presented in the form of frequency distribution. This gives us the information on the number of hours for which the velocity is within a specific range. For developing the frequency distribution, the wind speed domain is divided into equal intervals (say 0-1, 1-2, 2-3, etc.) and the number of times the wind record is within this interval is counted. An example for the monthly frequency distribution of wind speed is given in Table 3.2. If the velocity is presented in the form of frequency distribution, the average and standard deviation are given by

$$V_m = \left(\frac{\sum_{i=1}^{n} f_i \, V_i^{\,3}}{\sum_{i=1}^{n} f_i} \right)^{1/3} \quad \text{and} \tag{3.16}$$

$$\sigma_V = \sqrt{\frac{\sum_{i=1}^{n} f_i \, (V_i - V_m)^2}{\sum_{i=1}^{n} f_i}} \tag{3.17}$$

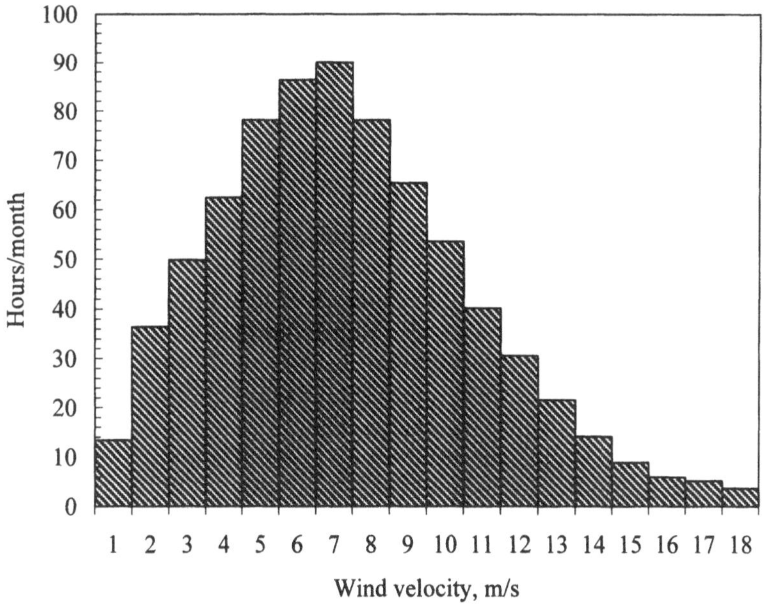

Fig. 3.16. Probability distribution of monthly wind velocity

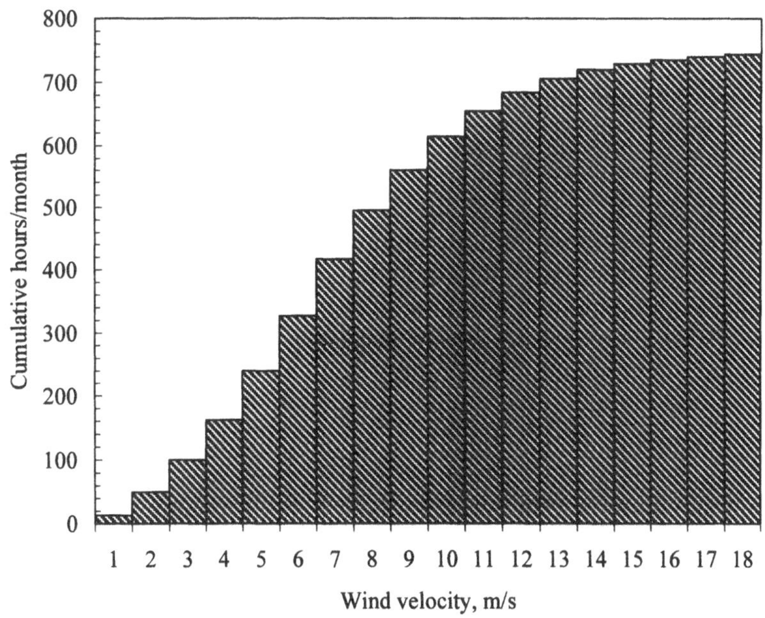

Fig. 3.17. Cumulative distribution of wind velocity

Here f_i is the frequency and V is the mid value of the corresponding interval. The average and standard deviation of wind data presented in Table 3.2 are 8.34 m/s and 0.81 m/s respectively. Frequency histogram based on the above data is shown in Fig. 3.16. It is interesting to note that, the average wind speed does not coincide with the most frequent wind speed. Generally, the average wind speed is higher than the most frequent wind speed, except for the trade winds which are relatively steady throughout the time. Cumulative distribution curve is constructed by plotting the cumulative time for which the wind velocity is below the upper limit of the wind class interval. The cumulative distribution of the above data is shown in Fig. 3.17.

3.3.3 Statistical models for wind data analysis

If we join the midpoints of the frequency and cumulative histograms in Fig. 3.16 and Fig. 3.17, we get smooth curves with a well defined pattern. This shows that it is logical to represent the wind velocity distributions by standard statistical functions. Various probability functions were fitted with the field data to identify suitable statistical distributions for representing wind regimes. It is found that the Weibull and Rayleigh distributions can be used to describe the wind variations in a regime with an acceptable accuracy level [7, 9, and 19].

3.3.3.1 Weibull distribution

Weibull distribution is a special case of Pierson class III distribution. In Weibull distribution, the variations in wind velocity are characterized by the two functions; (1) The probability density function and (2) The cumulative distribution function. The probability density function (f (V)) indicates the fraction of time (or probability) for which the wind is at a given velocity V. It is given by

$$f(V) = \frac{k}{c}\left(\frac{V}{c}\right)^{k-1} e^{-\left(V/c\right)^k} \tag{3.18}$$

Here, k is the Weibull shape factor and c is scale factor. The cumulative distribution function of the velocity V gives us the fraction of time (or probability) that the wind velocity is equal or lower than V. Thus the cumulative distribution F(V) is the integral of the probability density function. Thus,

$$F(V) = \int_0^{\alpha} f(V)\, dV = 1 - e^{-\left(V/c\right)^k} \tag{3.19}$$

Average wind velocity of a regime, following the Weibull distribution is given by

$$V_m = \int_0^{\infty} V\, f(V)\, dV \tag{3.20}$$

Substituting for f(V), we get

$$V_m = \int_0^{\infty} V\, \frac{k}{c}\left(\frac{V}{c}\right)^{k-1} e^{-\left(V/c\right)^k} dV \tag{3.21}$$

which can be rearranged as

$$V_m = k \int_0^\infty \left(\frac{V}{c}\right)^k e^{-\left(V/c\right)^k} \, dV \tag{3.22}$$

Taking

$$x = \left(\frac{V}{c}\right)^k, \; dV = \frac{c}{k} x^{\left(1/k - 1\right)} \, dx \tag{3.23}$$

Substituting for dV in Eq. (3.22),

$$V_m = c \int_0^\infty e^{-x} \, x^{1/k} \, dx \tag{3.24}$$

This is in the form of the standard gamma function, which is given by

$$\Gamma n = \int_0^\infty e^{-x} \, x^{n-1} \, dx \tag{3.25}$$

Hence, from Eq. (3.24), the average velocity can be expressed as

$$V_m = c \, \Gamma\left(1 + \frac{1}{k}\right) \tag{3.26}$$

The standard deviation of wind velocity, following the Weibull distribution is

$$\sigma_V = \left(\mu_2' - V_m^{\,2}\right)^{1/2} \tag{3.27}$$

Here, μ_2' is the second raw moment of the population which is given by

$$\mu_2' = \int_0^\infty V^2 \, f(V) \, dV \tag{3.28}$$

Substituting for $f(V)$ and from Eq. (3.23), we get

$$\mu_2' = c^2 \int_0^\infty e^{-X} \, x^{2/k} \, dx \tag{3.29}$$

Which can be expressed as a gamma integral in the form

$$\mu_2' = c^2 \, \Gamma\left(1 + \frac{2}{k}\right) \tag{3.30}$$

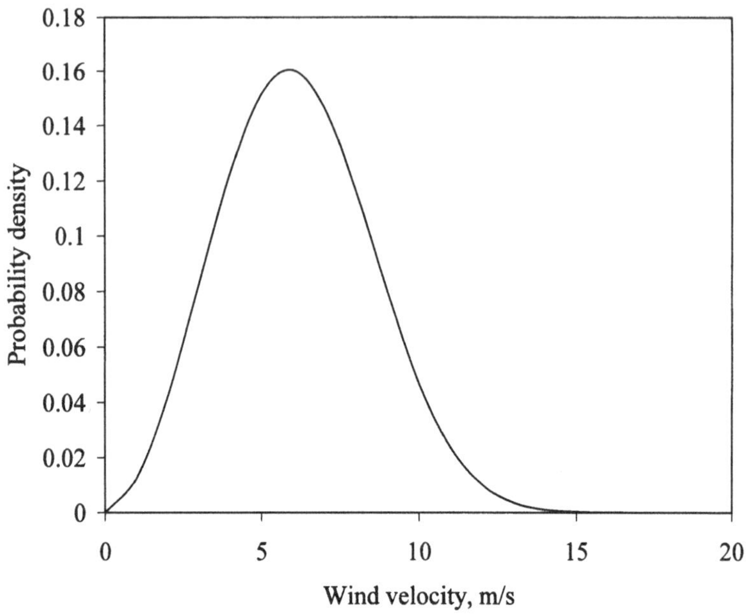

Fig. 3.18. Weibull probability density function

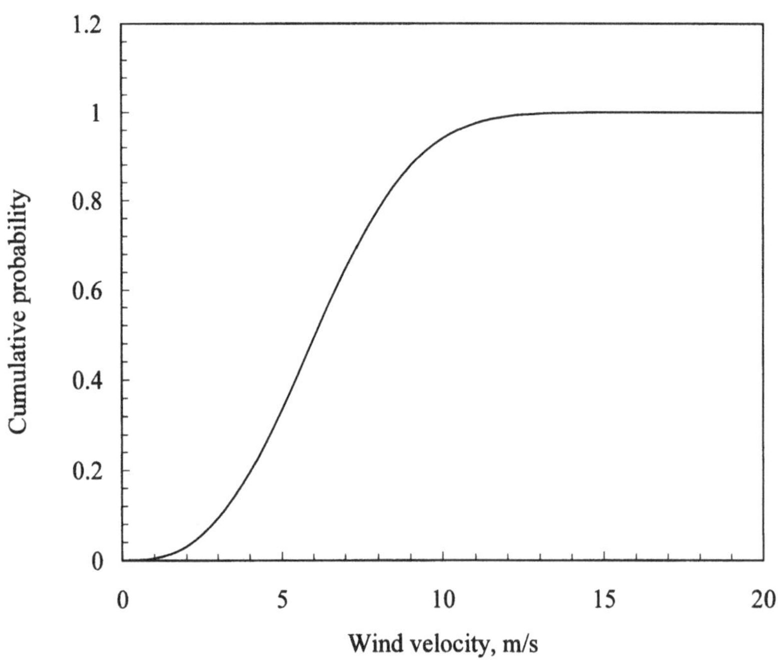

Fig. 3.19. Weibull cumulative distribution function

Substituting for μ_2' and V_m in Eq. (3.27),

$$\sigma_V = c\left[\Gamma\left(1+\frac{2}{k}\right) - \Gamma^2\left(1+\frac{1}{k}\right)\right]^{1/2} \tag{3.31}$$

The probability density and cumulative distribution functions of a wind regime, following the Weibull distribution are shown in Fig. 3.18 and 3.19. The k and c values for this site are 2.8 and 6.9 m/s respectively. The peak of the probability density curve indicates the most frequent wind velocity in the regime, which is 6 m/s in this case.

The cumulative distribution function can be used for estimating the time for which wind is within a certain velocity interval. Probability of wind velocity being between V_1 and V_2 is given by the difference of cumulative probabilities corresponding to V_2 and V_1. Thus

$$P(V_1 < V < V_2) = F\,(V_2) - F\,(V_1) \tag{3.32}$$

That is

$$P(V_1 < V < V_2) = e^{-\left(V_1/c\right)^k} - e^{-\left(V_2/c\right)^k} \tag{3.33}$$

We may be interested to know the possibilities of extreme wind at a potential location, so that the system can be designed to sustain the maximum probable loads. The probability for wind exceeding V_X in its velocity is given by

$$P\,(V > V_X) = 1 - \left(1 - e^{-\left(V_X/c\right)^k}\right) = e^{-\left(V_X/c\right)^k} \tag{3.34}$$

Example

A wind turbine with cut-in velocity 4 m/s and cut-out velocity 25 m/s is installed at a site with Weibull shape factor 2.4 and scale factor 9.8 m/s. For how many hours in a day, will the turbine generate power? Also estimate the probability of wind velocity to exceed 35 m/s at this site.

$$P(V_4 < V < V_{25}) = e^{-\left(4/9.8\right)^{2.4}} - e^{-\left(25/9.8\right)^{2.4}} = 0.89$$

Hence in a day, the turbine will generate power for $0.89 \times 24 = 21.36$ h

$$P\,(V > V_{35}) = e^{-\left(35/9.8\right)^{2.4}} = 0.000000001$$

Thus, the chance of wind getting stronger than 35 m/s at this site is very rare.

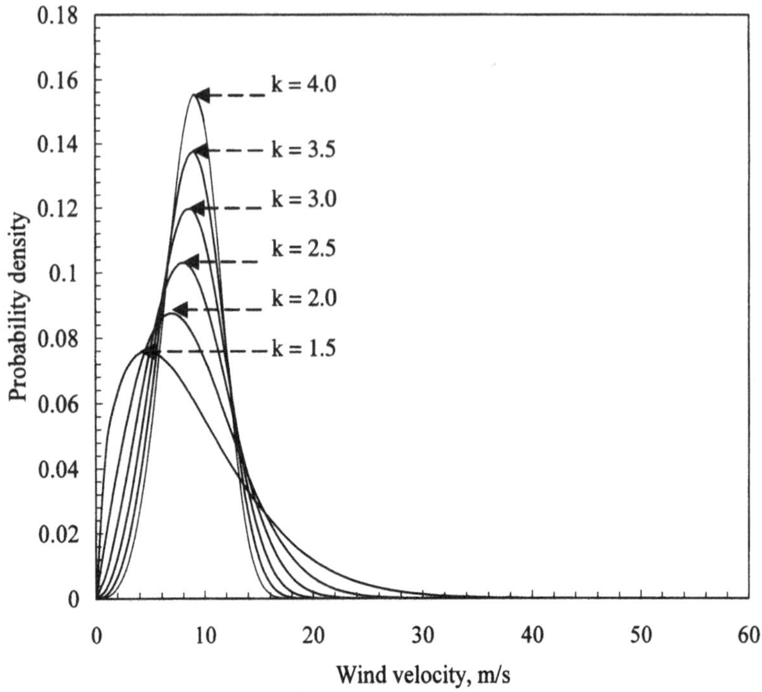

Fig. 3.20. Weibull probability density function for different shape factors

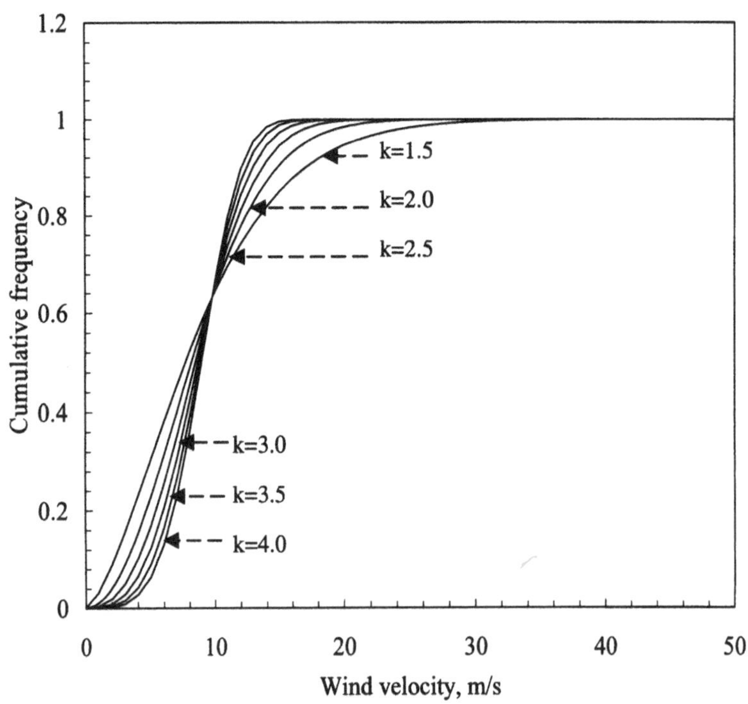

Fig. 3.21. Weibull cumulative distribution function for different shape factors

Under the Weibull distribution, the major factor determining the uniformity of wind is the shape factor k. Figures 3.20 and 3.21 illustrate the effect of k on the probability density and cumulative distribution functions of wind. Here, the scale factor is taken as 9.8 m/s. Uniformity of wind at the site increases with k. For example, with k =1.5, the wind velocity is between 0 and 20 m/s for 95 per cent of the time. Whereas, when k = 4, the velocity is more evenly distributed within a smaller range of 0 to 13 m/s for 95 per cent of the time. In the first case, the most frequent wind speed is 5 m/s, which is expected for 7.6 per cent of the total time. The most frequent wind speed of 9 m/s would prevail for 15.5 per cent of the time in the second case.

For analysing a wind regime following the Weibull distribution, we have to estimate the Weibull parameters k and c. The common methods for determining k and c are:

1. Graphical method
2. Standard deviation method
3. Moment method
4. Maximum likelihood method and
5. Energy pattern factor method

Graphical method

In the graphical method, we transform the cumulative distribution function in to a linear form, adopting logarithmic scales [5,18]. The expression for the cumulative distribution of wind velocity can be rewritten as

$$1 - F(V) = e^{-\left(V/c\right)^k} \tag{3.35}$$

Taking the logarithm twice, we get

$$\ln\{-\ln[1-F(V)]\} = k\ \ln(V_i) - k\ \ln c \tag{3.36}$$

Plotting the above relationship with ln (V_i) along the X axis and ln{-ln[1-F(V)]} along the Y axis, we get nearly a straight line. From Eq. (3.36), k gives the slope of this line and –k ln c represents the intercept. If we generate the regression equation for the plotted line using any standard spread sheets or statistical packages and compare it with Eq. (3.36), we can find out the values of k and c.

Example

The frequency distribution of wind velocity at a site is given in Table 3.3. Calculate the Weibull shape and scale factors.

First of all, we have to generate the cumulative distribution of the data based on the given frequencies. Each class should be represented by its upper limit as shown in the last column of the table. Now plot ln (V) in the X axis and ln {-ln [1-F(V)]} in the Y axis. The resulting graph is shown in Fig. 3.22. The points are slightly scattered. Fit a line through the points and deduce the best fit equation.

Table 3. 3. Frequency distribution of wind velocity

No	V (km/h)	Frequency	F(V)
1	0	0.002	0.002
2	1-2	0.005	0.007
3	3-4	0.008	0.015
4	5-6	0.014	0.029
5	7-8	0.025	0.054
6	9-10	0.037	0.091
7	11-12	0.048	0.139
8	13-14	0.051	0.19
9	15-16	0.057	0.247
10	17-18	0.051	0.298
11	19-20	0.069	0.367
12	21-22	0.07	0.437
13	23-24	0.073	0.51
14	25-26	0.074	0.584
15	27-28	0.072	0.656
16	29-30	0.066	0.722
17	31-32	0.058	0.78
18	33-34	0.054	0.834
19	35-36	0.041	0.875
20	37-38	0.033	0.908
21	39-40	0.028	0.936
22	41-42	0.021	0.957
23	43-44	0.017	0.974
24	45-46	0.011	0.985
25	47-48	0.008	0.993
26	49-50	0.004	0.997
27	51-52	0.002	0.999
28	53-54	0.001	1
29	55-56	0	1
30	57-58	0	1
31	59-60	0	1

The resulting equation for the plot is

$$y = 2.24\,x - 7.32 \tag{3.37}$$

The coefficient of determination (R^2) between the data and the fitted line is 0.98. Comparing Eq. (3.36) with Eq. (3.37), we get k for the given location as 2.24. Similarly, we have k ln c = 7.32. From this c can be solved as 26.31 km/h (7.31 m/s).

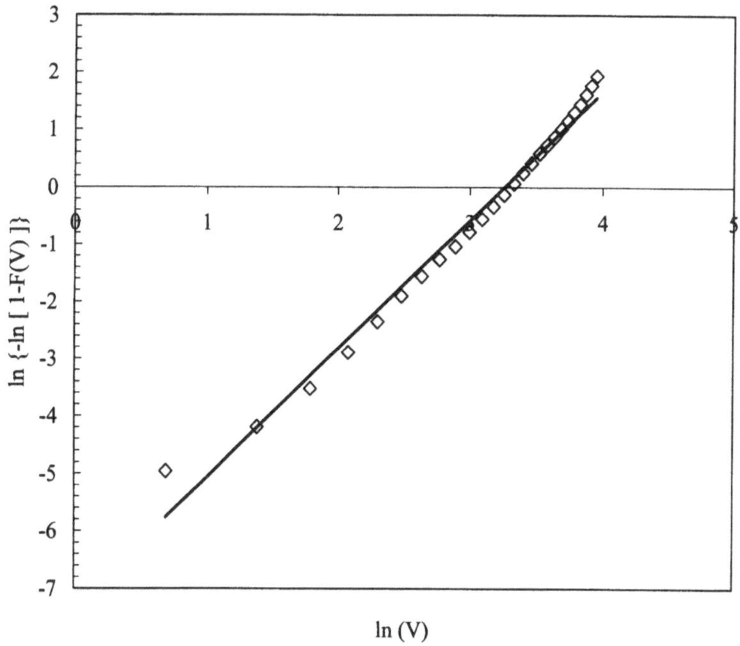

Fig. 3.22. Graphical method to estimate Weibull k and c

Standard deviation method

The Weibull factors k and c can also be estimated from the mean and standard deviation of wind data. Consider the expressions for average and standard deviation given in Eq. (3.26) and Eq. (3.31).
From these, we have

$$\left(\frac{\sigma_V}{V_m}\right)^2 = \frac{\Gamma\left(1+\frac{2}{k}\right)}{\Gamma^2\left(1+\frac{1}{k}\right)} - 1 \tag{3.38}$$

Once σ_V and V_m are calculated for a given data set, then k can be determined by solving the above expression numerically. Once k is determined, c is given by

$$c = \frac{V_m}{\Gamma\left(1+\frac{1}{k}\right)} \tag{3.39}$$

In a simpler approach, an acceptable approximation for k is [9, 2]

$$k = \left(\frac{\sigma_V}{V_m}\right)^{-1.090} \tag{3.40}$$

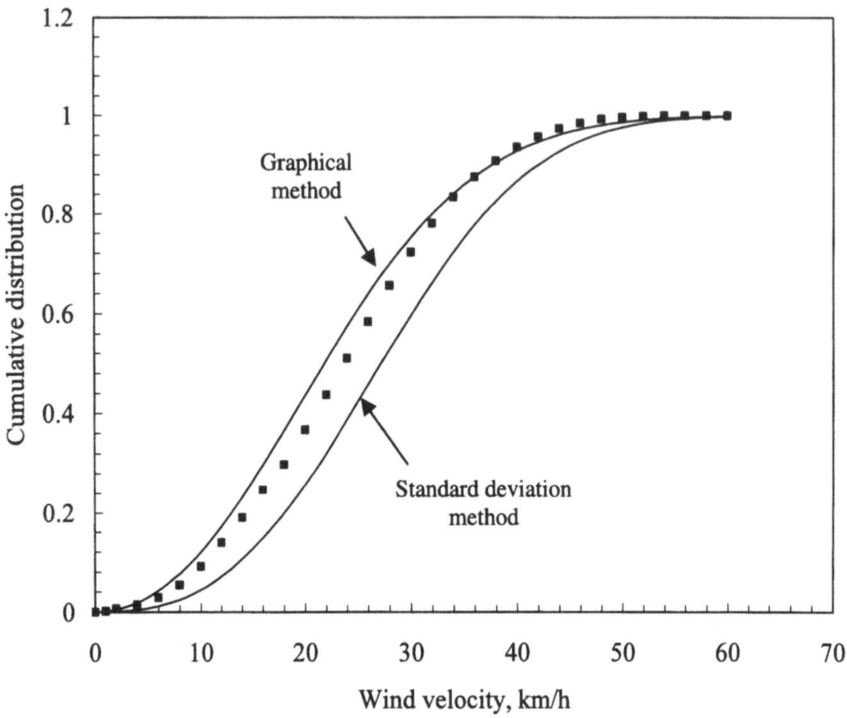

Fig. 3.23. Cumulative distribution generated using different methods

Similarly, c can be approximated as

$$c = \frac{2\,V_m}{\sqrt{\pi}} \tag{3.41}$$

More accurately, c can be found using the expression [1]

$$c = \frac{V_m\;k^{2.6674}}{0.184\;+\;0.816\;k^{2.73855}} \tag{3.42}$$

Example

Estimate the Weibull factors k and c for the data in Table 3.3 using the standard deviation method.

Mean and standard deviation of the given data, as per Eq. (3.16) and Eq. (3.17), are 28.08 km/h (7.80 m/s) and 10.88 km/h (3.02 m/s) respectively. Thus

$$k = \left(\frac{10.88}{28.08}\right)^{-1.090} = 2.81$$

Substituting for V_m and k in Eq. (3.42),

$$c = \frac{28.08 \times 2.81^{2.6674}}{0.184 + 0.816 \times 2.81^{2.73855}} = 31.6 \text{ km/h} = 8.78 \text{ m/s}$$

We can see that, there are differences in k and c estimated, following the graphical and standard deviation methods. The cumulative distributions generated using these methods are compared with the field measurements (represented by dotted lines) in Fig. 3.23. Results of the graphical method are found closer to the field observations. The standard deviation method also could predict the wind distribution within an acceptable level of accuracy.

Moment method

Another method for estimating k and c is the First and Second Order Moment Method. The nth moment M_n of Weibull distribution is given by

$$M_n = c^n \, \Gamma\left(1 + \frac{n}{k}\right) \tag{3.43}$$

If M_1 and M_2 are the first and second moments, with Eq. (3.43), c can be solved as

$$c = \frac{M_2}{M_1} \frac{\Gamma\left(1 + \frac{1}{k}\right)}{\Gamma\left(1 + \frac{2}{k}\right)} \tag{3.44}$$

Similarly,

$$\frac{M_2}{M_1^2} = \frac{\Gamma\left(1 + \frac{2}{k}\right)}{\Gamma^2\left(1 + \frac{1}{k}\right)} \tag{3.45}$$

In this method, M_1 and M_2 are calculated from the given wind data. Further, k and c are estimated by solving Eq. (3.44) and Eq. (3.45).

Maximum likelihood method

In the maximum likelihood method, the shape and scale factors are given as [3, 19]

$$k = \left[\frac{\sum_{i=1}^{n} V_i^k \ln(V_i)}{\sum_{i=1}^{n} V_i^k} - \frac{\sum_{i=1}^{n} \ln(V_i)}{n} \right]^{-1} \quad \text{and} \tag{3.46}$$

$$c = \left[\frac{1}{n}\sum_{i=1}^{n} V_i^k\right]^{1/k} \tag{3.47}$$

Energy pattern factor method

Energy pattern factor (E_{PF}) is the ratio between the total power available in the wind and the power corresponding to the cube of the mean wind speed [14]. Thus

$$E_{PF} = \frac{1/n \sum_{i=1}^{n} V_i^3}{\left(1/n \sum_{i=1}^{n} V_i\right)^3} \tag{3.48}$$

Once the energy pattern factor for a regime is found from the wind data, an approximate solution for k is

$$k = 3.957\ E_{PF}^{-0.898} \tag{3.49}$$

Various methods used for estimating k and c are discussed in detail in [19].

3.3.3.2 Rayleigh distribution

The reliability of Weibull distribution in wind regime analysis depends on the accuracy in estimating k and c. For the precise calculation of k and c, adequate wind data, collected over shorter time intervals are essential. In many cases, such information may not be readily available. The existing data may be in the form of the mean wind velocity over a given time period (for example daily, monthly or yearly mean wind velocity). Under such situations, a simplified case of the Weibull model can be derived, approximating k as 2. This is known as the Rayleigh distribution. Taking $k = 2$ in Eq. (3.26) we get

$$V_m = c\,\Gamma\left(3/2\right) \tag{3.50}$$

Evaluating the above expression and rearranging,

$$c = \frac{2V_m}{\sqrt{\pi}} \tag{3.51}$$

Substituting for c in Eq. (3.18), we get

$$f(V) = \frac{\pi}{2}\frac{V}{V_m^2}\ e^{-\left[\pi/4\left(V/V_m\right)^2\right]} \tag{3.52}$$

Similarly, the cumulative distribution is given by

$$F(V) = 1 - e^{-\left[\frac{\pi}{4}\left(\frac{V}{V_m}\right)^2\right]} \tag{3.53}$$

Thus we can calibrate the probability density and cumulative distribution functions of wind on the basis of its average velocity. The validity of Rayleigh distribution in wind energy analysis has been established by comparing the Rayleigh generated wind pattern with long term field data [4]. Under the Rayleigh distribution, the probability of wind velocity to be between V_1 and V_2 is

$$P(V_1 < V < V_2) = e^{-\frac{\pi}{4}\left(\frac{V_1}{V_m}\right)^2} - e^{-\frac{\pi}{4}\left(\frac{V_2}{V_m}\right)^2} \tag{3.54}$$

The probability of wind to exceed a velocity of V_X is given by

$$P(V > V_X) = 1 - \left(1 - e^{-\frac{\pi}{4}\left(\frac{V_X}{V_m}\right)^2}\right) = e^{-\frac{\pi}{4}\left(\frac{V_X}{V_m}\right)^2} \tag{3.55}$$

Example

Generate the cumulative distribution for the wind data in Table 3.3 using the Rayleigh distribution. Compare the result with the Weibull model and field data. For the given data, the mean and standard deviation of wind velocity are 28.08 km/h and 10.88 km/h respectively. Weibull k and c for the location are 2.24

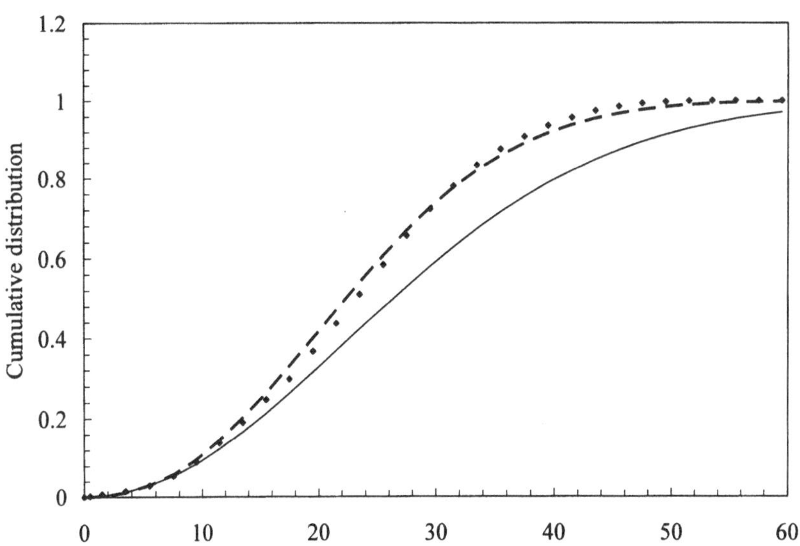

Fig. 3.24. Comparison of field data with Weibull and Rayleigh

and 26.31 km/h. The cumulative distributions are compared in Fig. 3.24. Scattered points show the field data, whereas dashed and continuous curves represent the Weibull and Rayleigh models respectively. The wind distribution simulated using the Weibull model is very close to field observations

Other distributions are also being used for wind regime analysis. Examples are Gamma distribution [15], Log normal distribution [5] and Logistic distribution [6]. However, the Weibull and Rayleigh models are widely accepted and extensively used for wind regime analysis. Hence we will focus our further discussions on these two distributions.

3.4 Energy estimation of wind regimes

Assessing the energy available in the wind regime prevailing at a site is one of the preliminary steps in the planning of a wind energy project. Wind energy density and the energy available in the regime over a period are usually taken as the yardsticks for evaluating the energy potential. The wind energy density (E_D) is the energy available in the regime for a unit rotor area and time. Thus, E_D is a function of the velocity and distribution of wind in the regime. We can arrive at the total energy available in the spectra (E_S), by multiplying the wind energy density by the time factor.

Other factors of interest are the most frequent wind velocity ($V_{F\,max}$) and the velocity contributing the maximum energy ($V_{E\,max}$) to the regime. Peak of the probability density curve represents $V_{F\,max}$. Due to the cubic velocity-power relationship of wind, the velocity contributing the maximum energy is usually higher than the most frequent wind velocity. A WECS operates at its maximum efficiency at its design wind velocity, V_d. Hence, it is advantageous that V_d and $V_{E\,max}$ are made as close as possible. Once the velocity responsible for maximum energy could be identified for a particular site, the designer can design his system to be most efficient at this velocity or more practically, a planner can select a machine having design wind speed very close to V_{Emax} , if he is not constrained by other factors. In the following sections we will discuss the methods to estimate the energy potential of a wind regime based on the above indices, considering both the Weibull and Rayleigh models.

3.4.1 Weibull based approach

For a unit area of the rotor, power available (P_V) in the wind stream of velocity V is

$$P_V = \frac{1}{2}\,\rho_a\,V^3 \tag{3.56}$$

The fraction of time for which this velocity V prevails in the regime is given by f(V). The energy per unit time contributed by V is PV f(V). Thus the total energy, contributed by all possible velocities in the wind regime, available for unit rotor area and time may be expressed as

$$E_D = \int_0^{\infty} P_V \; f(V) \, dV \tag{3.57}$$

Substituting for P(V) and f(V) in the above expression and simplifying, we get

$$E_D = \frac{\rho_a k}{2 \, c^k} \int_0^{\infty} V^{(k+2)} \; e^{-\left(V/c\right)^k} \, dV \tag{3.58}$$

Now taking

$$x = \left(V/c\right)^k \tag{3.59}$$

We can rewrite the above expression as

$$E_D = \frac{\rho_a \, c^3}{2} \int_0^{\infty} x^{3/k} \; e^{-x} \, dx \tag{3.60}$$

Comparing the above expression with the standard gamma integral, we get

$$E_D = \frac{\rho_a \, c^3}{2} \; \Gamma\left(\frac{3}{k} + 1\right) \tag{3.61}$$

Applying the reduction formula

$$\Gamma(n) = (n-1)\,\Gamma(n-1) \tag{3.62}$$

we get energy density as

$$E_D = \frac{\rho_a \, c^3}{2} \; \frac{3}{k} \; \Gamma\left(\frac{3}{k}\right) \tag{3.63}$$

Once E_D is known, energy available over a period (E_I) can be calculated as

$$E_I = E_D \, T = \frac{\rho_a \, c^3 T}{2} \left(\frac{3}{k}\right) \Gamma\left(\frac{3}{k}\right) \tag{3.64}$$

where T is the time period. For example, T is taken as 24 when we calculate the energy on daily basis.

Referring to Fig. 3.16, peak of the probability density curve represents the most frequent wind velocity V_F. For our analysis, let us rewrite the probability density function as

$$f(V) = \frac{k}{c^k} \; V^{k-1} \; e^{-\left(V/c\right)^k} \tag{3.65}$$

For the most frequent wind velocity,

$$f'(V) = 0 \tag{3.66}$$

which means that,

$$\frac{k}{c^k}\; e^{-\left(V/c\right)^k} \left[-\frac{k}{c^k}\, V^{2(k-1)} + (k-1)\, V^{(k-2)} \right] = 0 \tag{3.67}$$

Solving the above expression for V, we get

$$V = c\left(\frac{k-1}{k}\right)^{1/k} \tag{3.68}$$

It can also be observed that

$$f'(V) > 0 \text{ in the interval} \left[0,\; c\left(\frac{k-1}{k}\right)^{1/k}\right] \tag{3.69}$$

and

$$f'(V) < 0 \text{ in the interval} \left[c\left(\frac{k-1}{k}\right)^{1/k},\; \infty\right) \tag{3.70}$$

This implies that f(V) is maximum at V represented by Eq. (3.68). We will indicate the most frequent wind velocity in the regime by $V_{F\,Max}$. Thus,

$$V_{F\,Max} = c\left(\frac{k-1}{k}\right)^{1/k} \tag{3.71}$$

Now let us consider the velocity contributing maximum energy to the regime (V_{EMax}). Energy per unit rotor area and time, contributed by a velocity V is

$$E_V = P_V\; f(V) \tag{3.72}$$

Substituting for P_V and f(V), we get

$$E_V = \frac{\rho_a V^3}{2}\; \frac{k}{c}\left(\frac{V}{c}\right)^{k-1} e^{-\left(V/c\right)^k} \tag{3.73}$$

Introducing a term B such that

$$B = \frac{\rho}{2}\; \frac{k}{c^k} \tag{3.74}$$

we get

$$E_V = B\, V^{(k+2)}\, e^{-\left(V/c\right)^k} \tag{3.75}$$

For E_V to be maximum,

$$E_V' = 0 \tag{3.76}$$

This implies that

$$B\left(- e^{-\left(V/c\right)^k} V^{k+2} \frac{k}{c^k} V^{(k-1)} + e^{-\left(V/c\right)^k} (k+2)\, V^{(k+1)} \right) = 0 \tag{3.77}$$

Solving the above expression for V, we get

$$V = \frac{c\ (k+2)^{1/k}}{k^{1/k}} \tag{3.78}$$

It can be observed that the function E_V is increasing in the interval

$$\left[0, \frac{c\ (k+2)^{1/k}}{k^{1/k}} \right] \tag{3.79}$$

and decreasing in the interval

$$\left[\frac{c\ (k+2)^{1/k}}{k^{1/k}}, \infty \right) \tag{3.80}$$

Hence, V represented by Eq. (3.78) gives us the velocity contributing maximum energy to the regime. Representing this by $V_{E\,Max}$, we have

$$V_{E\,Max} = \frac{c\ (k+2)^{1/k}}{k^{1/k}} \tag{3.81}$$

The WERA programme

As we can see, the above expressions deduced for the energy analysis cannot be resolved analytically. Numerical methods are to be adopted for solving these equations. For readers, who are not familiar with numerical methods and high-end programming, a software named WERA has been developed. Present version of WERA, provided with the book, has six modules arranged in three menus as

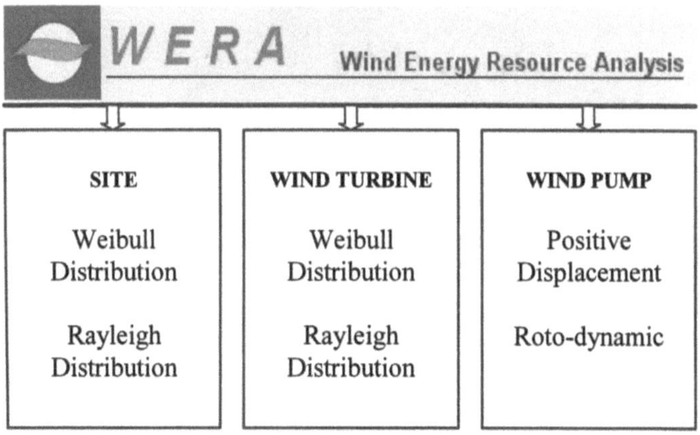

Fig. 3.25. The WERA programme

shown in Fig. 3.25. The programme is designed to be simple and friendly to use. More detailed of WERA are given in Appendix 1.

Example

Estimate the wind energy density, annual energy intensity, most frequent wind velocity and the velocity responsible for maximum energy for the wind regime given in Table. 3.3.
k and c for this site are 2.24 and 7.31 m/s respectively. WEIBULL analysis under the SITE module of the WERA programme is used to solve this problem. T is taken as 365 X 24 = 8760 h. The results are given below.

The most frequent wind velocity	=	5.61 m/s
Velocity contributing the maximum energy	=	9.72 m/s
Energy density	=	0.287 kW/m^2
Annual energy intensity	=	2515.19 kWh/m^2

3.4.2 Rayleigh based approach

Considering Rayleigh distribution, wind energy density can be expressed as

$$E_D = \int_0^\infty P_V \; f(V) \; dV = \int_0^\infty \frac{\pi \, \rho_a}{4 \, V_m^2} \; V^4 \; e^{-\left[\frac{\pi}{4}\left(\frac{V}{V_m}\right)^2\right]} \; dV \tag{3.82}$$

Introducing a constant K such that

$$K = \frac{\pi}{4 \, V_m^2}, \tag{3.83}$$

the energy density**Error! Bookmark not defined.** is given by

$$E_D = K\ \rho_a \int_0^{\infty} V^4\ e^{\left(-KV^2\right)}\ dV \tag{3.84}$$

Now let us substitute

$$x = K\,V^2 \tag{3.85}$$

Hence,

$$dV = \frac{dx}{2\sqrt{Kx}} \tag{3.86}$$

With this substitution, E_D can be expressed as

$$E_D = \frac{\rho_a}{2\ K^{3/2}} \int_0^{\infty} x^{3/2}\ e^{-x}\ dx \tag{3.87}$$

The above expression can be reduced to a gamma function in the form

$$E_D = \frac{\rho_a}{2\ K^{5/2}}\ \Gamma\left(\frac{5}{2}\right) \tag{3.88}$$

which can be further evaluated as

$$E_D = \frac{3}{8}\ \frac{\rho_a\ \sqrt{\pi}}{K^{1.5}} \tag{3.89}$$

Now, replacing K, we get the energy density at the site as

$$E_D = \frac{3}{\pi}\ \rho_a\ V_m^{\ 3} \tag{3.90}$$

From E_D, energy available for the unit area of the rotor, over a time period, can be estimated using the expression

$$E_I = T\ E_D \ = \ \frac{3}{\pi} T\ \rho_a\ V_m^{\ 3} \tag{3.91}$$

For identifying the most frequent wind speed, the probability density function of the Rayleigh distribution is rewritten in terms of the constant K. Thus,

$$f(V) \ = \ 2\,K\,V\ e^{-\left(KV^2\right)} \tag{3.92}$$

Now, the condition of maxima is applied to the above expression. This yields

$$f'(V) = 0 \tag{3.93}$$

which implies that

$$2K\ e^{-\left(KV^2\right)}\left(1-2KV^2\right)=0 \tag{3.94}$$

Solving Eq. (3.94) for V, we get

$$V=\frac{1}{\sqrt{2K}} \tag{3.95}$$

It can also be observed that

$$f'(V)>0 \text{ in the interval } \left[0, \frac{1}{\sqrt{2K}}\right) \tag{3.96}$$

and

$$f'(V)<0 \text{ in the interval } \left[\frac{1}{\sqrt{2K}}, \infty\right) \tag{3.97}$$

Hence, f(V) is maximum at V represented by Eq. (3.95). Let us represent this by $V_{F\,Max}$. Thus we have

$$V_{F\,Max}=\frac{1}{\sqrt{2K}}=\sqrt{\frac{2}{\pi}}\ V_m \tag{3.98}$$

Now, let us identify the velocity contributing maximum energy to the regime. Following the Rayleigh model, the energy corresponding to a wind velocity V, over unit rotor area and time can be expressed as

$$E_V=P_V\ f(V)=K\ \rho_a\ V^4\ e^{\left(-KV^2\right)} \tag{3.99}$$

For maximum value of E_V,

$$E_V'=0 \tag{3.100}$$

This implies that

$$K\ \rho_a\ e^{-KV^2}\left(4\,V^3+V^4\left(-2KV\right)\right)=0 \tag{3.101}$$

Solving the above expression for V, we get

$$V=\sqrt{\frac{2}{K}} \tag{3.102}$$

Here we note that E_V is increasing in the interval

$$\left[0, \sqrt{\frac{2}{K}}\right] \tag{3.103}$$

and decreasing in the interval

$$\left[\sqrt{\frac{2}{K}},\ \infty\right) \tag{3.104}$$

Hence, E_V has the maximum value for V represented by Eq. (3.102). Thus

$$V_{E\,Max} = \sqrt{\frac{2}{K}} = 2\sqrt{\frac{2}{\pi}}\ V_m \tag{3.105}$$

In contrast with the Weibull based analysis, Rayleigh based analysis could yield simple expressions, which can be solved analytically.

Example

Monthly wind velocity at a location is given in Table. 3.4. Calculate the wind energy density, monthly energy availability, most frequent wind velocity and the velocity corresponding to the maximum energy.

Table 3.4. Monthly average wind velocity

Jan	Feb	Mar	Apr	May	Jun	Jul	Aug	Sep	Oct	Nov	Dec
9.14	8.3	7.38	7.29	10.1	11.1	11.4	11.1	10.3	7.11	6.74	8.58

Let us use WERA for solving this problem. RAYLEIGH analysis under the SITE module is used here. Results are shown in Table 3.5.

Table 3.5. Wind energy potential based on the Rayleigh analysis

Month	E_D (kW/m^2)	E_I (kW/m^2/month)	V_{Fmax} (m/s)	V_E (m/s)
January	0.90	666.95	7.29	14.58
February	0.67	451.11	6.62	13.24
March	0.47	351.09	5.89	11.77
April	0.46	327.49	5.82	11.63
May	1.20	889.30	8.03	16.05
June	1.59	1146.72	8.83	17.66
July	1.76	1307.78	9.13	18.25
August	1.59	1184.94	8.83	17.66
September	1.29	931.78	8.24	16.48
October	0.42	313.95	5.67	11.34
November	0.36	258.82	5.38	10.75
December	0.74	551.72	6.84	13.69

References

1. Ananstasios B Dimitrios C Thodoris DK (2002) A nomogram method for estimating the energy produced by wind turbine generators. Solar Energy 72(3): 251-259
2. Boweden GJ Barker PR Shestopal VO Twidell JW (1983) The Weibull distribution function and wind statistics. Wind Engineering 7:85-98
3. Chang TJ Wu YT Hsu HY Chu CR Liao CM (2003) Assessment of wind characteristics and wind turbine characteristics in Taiwan. Renewable Energy 28:851-871
4. Corotis RB Sigl AB Klein J (1978) Probability models for wind energy magnitude and persistence. Solar Energy 20:483-493
5. Garcia A Torres JL Prieto E DE Francisco A (1998) Fitting wind speed distributions: a case study. Solar Energy 62(2): 139-144
6. Guzzi R Justus CG (1988) Physical climatology of solar and wind energy. World Scientific, Singapore
7. Hennessey JP (1977) Some aspects of wind power statistics. J Applied Meteorology 16: 119-128
8. Hewson EW Wade JW (1977) Biological wind prospecting. Third Wind Energy workshop (CONF 770921), Washington DC, pp 335-348
9. Justus CG Hargraves WR Mikhail A Graber D (1978) Methods of estimating wind speed frequency distribution. J Applied Meteorology 17: 350-353
10. Lysen EH (1983) Introduction to wind energy. CWD, Amersfoort, Netherlands
11. Machias AV, Skikos, GD (1992) Fuzzy risk index for wind sites. IEEE Transactions on Energy Conversion 7(4): 638-643
12. Marrs RW Marwitz J (1977) Locating areas of high wind energy potential by remote observation of eolian geomorphology and topography. Third Wind Energy workshop (CONF 770921), Washington DC, pp307-320
13. Mathew S, Pandey KP, Anil KV (2002) Analysis of wind regimes for energy estimation. Renewable Energy 25:381-399
14. Mengelkamp HT (1988) On the energy output estimation of wind turbines. J Energy Research 12:113-123
15. Panda RK Sarkar TK Bhattacharya AK (1990) Stochastic study of wind energy potential in India. Energy 15(10): 921-930
16. Parsa MJS Mapdi M (1995) Wind power statistics and evaluation of wind power density. Renewable Energy: 6(5/6) 623-628
17. Putnam PC (1948) Power from the wind. Van Nostrand, New York
18. Scerri E Farrugia R (1996) Wind data evaluation in the Maltese Islands. Renewable Energy 7(1): 109-114
19. Stevens MJM Smulders PT (1979) The estimation of parameters of the Weibull wind speed distribution for wind energy utilization purposes. Wind Engineering 3(2): 132-145
20. http:\\www.WINDPOWER.org

4 Wind energy conversion systems

During its transition from the earlier day's wind 'mills' to the modern wind electric generators, the wind energy conversion systems (WECS) have transformed to various sizes, shapes and designs, to suit the applications for which they are intended for. For example, at the inceptive stage of the technology, wind machines were used for grinding grains. Hence these ancient machines had vertical axis with wind catching surfaces made of canvas or bundles of reeds. The mechanical power available at the shaft was utilized for grain milling. With the advent of technology, so-called 'American wind mills' were introduced in the 19th century. These systems were designed with multi bladed rotor, mechanically coupled with reciprocating piston pumps, which was appropriate for water pumping application. The era of wind electric generators started in 1890 with the construction of the turbine in Denmark for meeting the rural electricity demand.

The modern wind turbine is a sophisticated piece of machinery with aerodynamically designed rotor and efficient power generation, transmission and regulation components. Size of these turbines ranges from a few Watts to several Mega Watts. Modern trend in the wind industry is to go for bigger units of several MW capacities, as the system scaling up can reduce the unit cost of wind generated electricity. Most of today's commercial machines are horizontal axis wind turbines (HAWT) with three bladed rotors. Though research and development activities on vertical axis wind turbines (VAWT) were intense during the end of the last century, VAWT could not evolve as a reliable alternative to the horizontal axis machines.

The turbines may be grouped into arrays, feeding power to a utility, with its own transformers, transmission lines and substations. Stand-alone systems catering the needs of smaller communities are also common. As wind is an intermittent source of energy, hybrid systems with back up from diesel generators or photovoltaic panels are also popular in remote areas.

For the efficient and reliable performance of a WECS, all its components are to be carefully designed, crafted and integrated. In this chapter, we will discuss the constructional features of WECS giving emphasis to various components, systems and sub-systems. As off shore installations are getting prominence in the recent years, details of such turbines are also included. Wind powered water pumps, which are still relevant in remote rural areas, are also featured in this chapter.

4.1 Wind electric generators

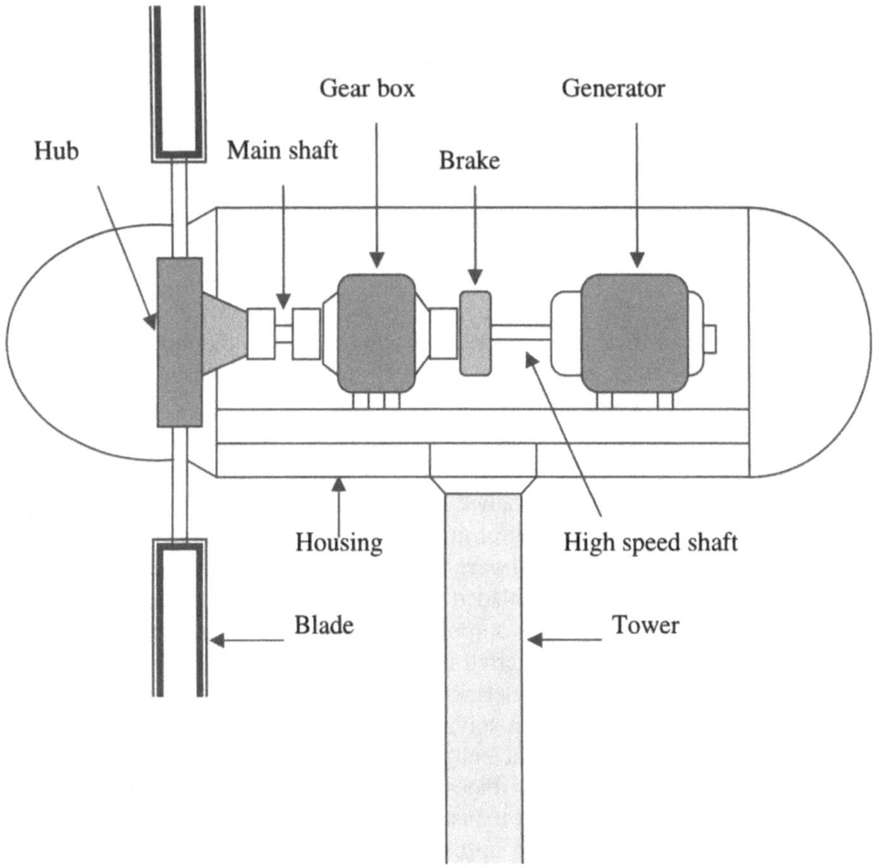

Fig. 4.1. Components of a wind electric generator

Electricity generation is the most important application of wind energy today. The major components of a commercial wind turbine are:

1. Tower
2. Rotor
3. High speed and low speed shafts
4. Gear box
5. Generator
6. Sensors and yaw drive
7. Power regulation and controlling units
8. Safety systems

The major components of the turbine are shown in Fig. 4.1.

4.1.1 Tower

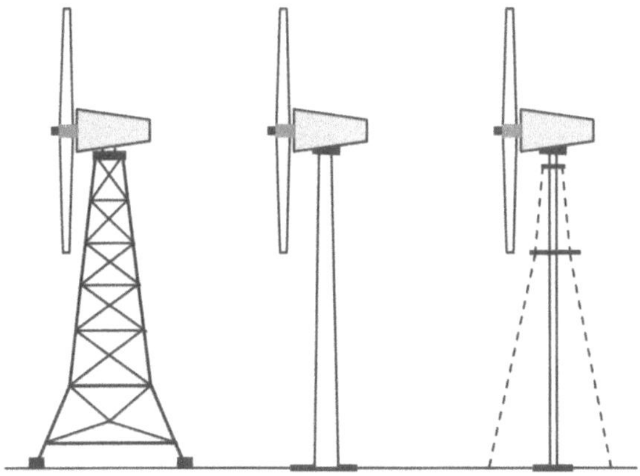

Lattice tower Tubular tower Guyed tower

Fig. 4.2. Different types of towers

Tower supports the rotor and nacelle of a wind turbine at the desired height. The major types of towers used in modern turbines are lattice tower, tubular steel tower and guyed tower. Schematic views of these towers are shown in Fig. 4.2.

The lattice towers are fabricated with steel bars joined together to form the structure as shown in the figure. They are similar to the transmission towers of electric utilities. Lattice towers consume only half of the material that is required for a similar tubular tower. This makes them light and thus cheaper. For example, lattice tower for a typical turbine may cost $ 25,000 less than the tubular option of similar size [10]. Legs of these towers are spread widely as shown in the figure. As the load is distributed over a wider area, these towers require comparatively lighter foundation, which will again contribute to the cost reduction.

Lattice towers have several demerits. The major problem is the poor aesthetics as they may be visually unacceptable to some viewers. Similarly, avian activities are more intense around the lattice towers as the birds can conveniently perch on its horizontal bars. This may increase the rate of avian mortality as discussed in Chapter 6. Lattice towers are not maintenance friendly. At installations where the temperature drops to an extremely low level, maintenance of systems with lattice towers is difficult as workers would be exposed to the chilling weather. Moreover, as these towers do not have any lockable doors, they are less secure for maintenance.

Due to these limitations, most of the recent installations are provided with tubular steel towers. These towers are fabricated by joining tubular sections of 10 to 20 m length. The complete tower can be assembled at the site within 2 or 3 days. The tubular tower, with its circular cross-section, can offer optimum bending resistance in all directions. These towers are aesthetically acceptable and pose less danger to the avian population.

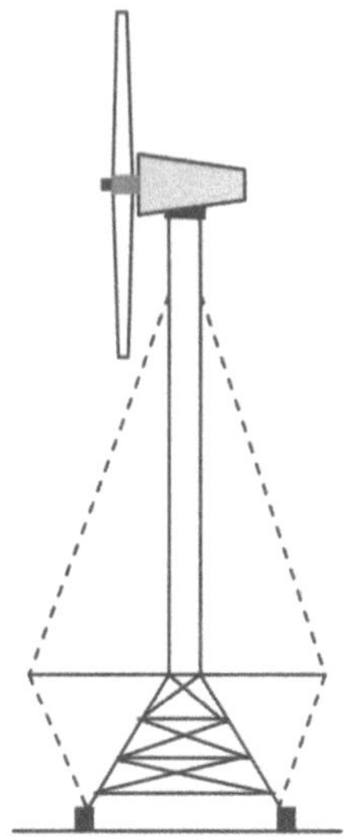

Fig. 4.3. The hybrid tower concept

For small systems, towers with guyed steel poles are being used. By partially supporting the turbine on guy wires, weight and thus the cost of the tower can be considerably reduced. Usually, four cables equally spaced and inclined at 45°, support the tower. As accesses to these towers are difficult, they are not popular with large scale installations. However, guyed towers are in use even with MW sized installations in Germany [2].

Load acting on the tower increases with size of the turbine. Hence, the recent trend for MW sized systems would in turn demand for higher tower dimensions in terms of diameter and wall thickness. This impose limitations while transportation of these towers. Usually, inland transportation of structures with size higher than 4.3 m and weight more than 50 to 60 tons is difficult. Further, fabricating these huge structures is not an easy task, as rolling and welding plates with wall thickness more than 50 mm is difficult. Due to these problems, hybrid towers are proposed for high capacity systems.

In concrete-tubular hybrid tower, the lower part is made of concrete where as the upper part is with conventional tubular steel structure. One of such design uses prefabricated long and narrow concrete elements for the lower portion [2]. These elements are 10-15 m long, 3-4 m wide and 250-350 mm thick. With these dimensions, they can be transported in a truck. Steel cables, protected against corrosion, are used to pre-stress the concrete in this design. The whole structure can be assembled at the field within a week.

Another design of the hybrid towers is proposed by NREL [15]. This is a combination of truss, tubular and guyed towers, as shown in Fig. 4.3. The lower truss portion allows the design to utilize the advantage of larger foot prints, whereas with the guyed wires, size of tubular steel can be reduced. Cost of this tower is

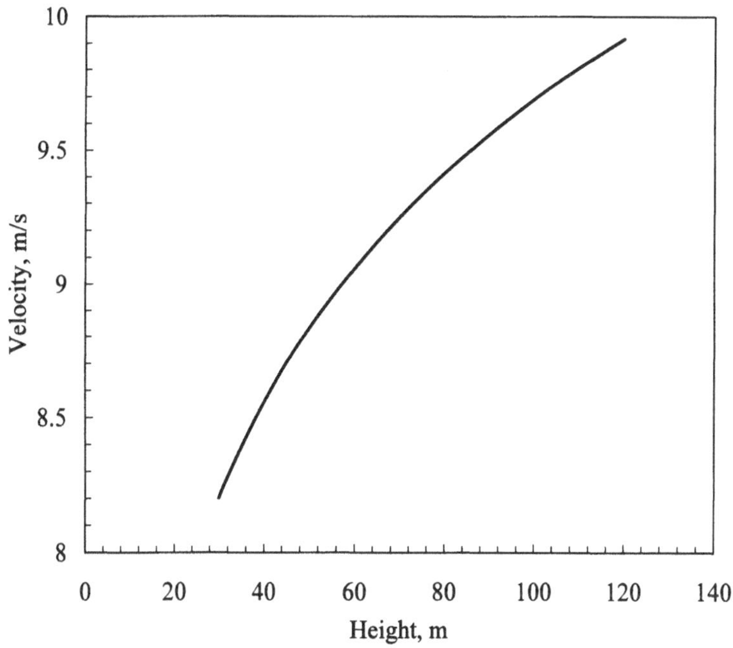

Fig. 4.4. Effect of tower height on the velocity at hub height

slightly higher than other tower options. However, this concept could reduce the maximum tube diameter required, resolving the transport problem. Similarly, the total mass of the structure could be reduced by 25 per cent.

Wind velocity increases with height due to wind shear. Hence, the taller the tower, the higher will be the power available to the rotor. Rate at which the available power increases with height depends on the surface roughness of the ground. Effect of wind shear on vertical wind profile has already been discussed in Chapter 3. The ratio of velocities at two heights Z_R and Z is given by,

$$\frac{V(Z_R)}{V(Z)} = \frac{\ln\left(Z_R/Z_0\right)}{\ln\left(Z/Z_0\right)} \tag{4.1}$$

Here Z_0 is the roughness height of the terrain.

Using the above expression, let us examine the effect of tower height on the performance of a typical wind turbine. Suppose the average wind velocity at the site, at 30 m tower height, is 8.2 m/s. Let the roughness height be 0.04 m, which is logical for smooth open grass lands. Change in velocity with height under these conditions is shown in Fig. 4.4. The velocity at 120 m hub height is 9.9 m/s, which means that, the available power is 1.75 times higher than that at 30 m height. In tune with the increase in velocity, capacity factor (defined in chapter 5) of the turbine also improves as shown in Fig. 4.5. For example, the capacity factor at 30 m is 0.26 whereas, at 120 m height, the system could attain a capacity factor of 0.37.

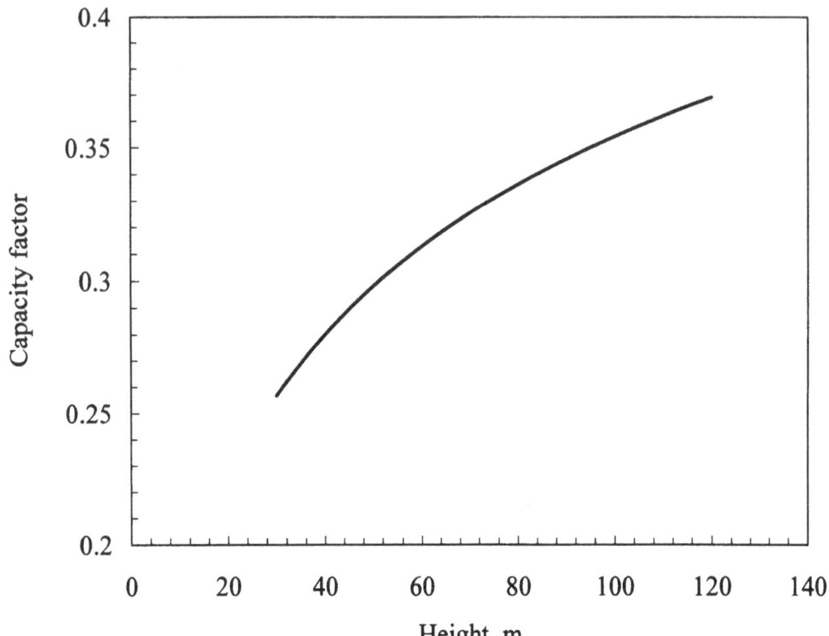

Fig. 4.5. Effect of tower height on the capacity factor

Apart from the increase in wind velocity, better matching between the wind spectra and the turbine also improves the capacity factor. This aspect will be discussed later in Chapter 5.

Thus, performance of a wind turbine improves with its tower height. However, taller towers cost more. Towers account for around 20 per cent of the total systems cost. At present, cost of every additional 10 m of tower is approximately $15,000. This means that, while we increase the tower height from 30 m to 120 m as in the example, the system cost would shoot up by $ 135,000. Can this additional cost be justified by the improvement in system performance? In other words, what should be the optimum tower height? The best criterion is to work out the economics of the options in terms of cost per kWh of electricity generated.

In Chapter 7, we will derive that the cost of wind generated electricity (c) is given by

$$c = \frac{C_I}{8760\,n}\left(\frac{1}{P_R\,C_F}\right)\left\{1+m\left[\frac{(1+I)^n-1}{I\,(1+I)^n}\right]\right\} \tag{4.2}$$

where C_I is the capital investment for the system, n is the expected life, P_R is the rated power, C_F is the capacity factor, m is the maintenance cost and I is the discounting rate corrected for inflation and escalation. Here, C_I increases with the tower height. However, with increase in height, C_F also improves. Thus, the selection of the optimum tower height is ultimately a trade-in between the cost of the system and its performance.

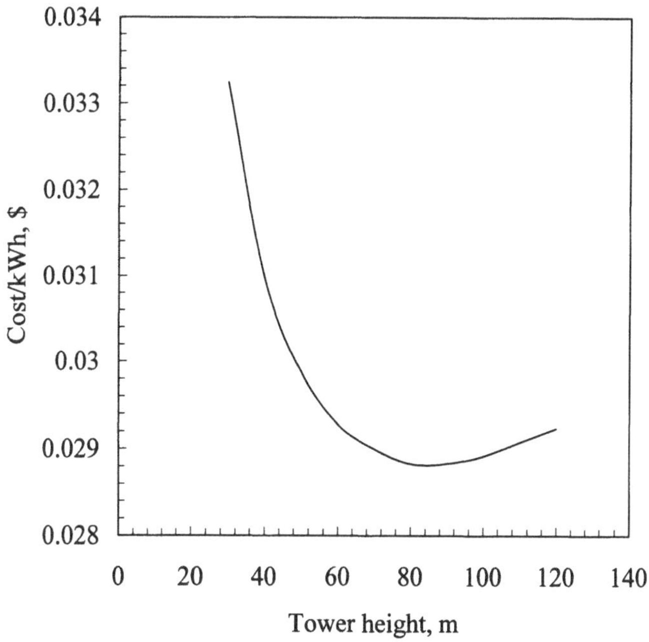

Fig. 4.6. Effect of tower height on generation cost

Example

Cost of a 600 kW wind turbine and attachments, with the standard 30 m tower, is $ 650,000. Every 10 m tower cost an additional $ 15,000. Expected life of the turbine is 20 years. Annual operational and maintenance costs accounts for 3.5 per cent of the initial cost. Discounting rate, corrected for inflation and escalation is 7 per cent. Average wind velocity at the site is 8.2 m/s and the roughness height is 0.04 m. Estimate the optimum tower height for the system.

Here, first of all, we have to compute the wind velocities at different tower heights using Eq. (4.1). The capacity factor of the system, corresponding to these velocities, can be computed using the 'WIND TURBINE-RAYLEIGH' module of WERA programme. Once the capacity factors for different tower heights are available, corresponding unit cost of electricity can be estimated using Eq. 4.2. It should be noted that, for increase in every 10 m tower height, $ 15,000 should be added to the capital cost.

The results are plotted in Fig. 4.6. Upto a tower height of 80 m, the cost of kWh generated is declining gradually due to the improvement in the turbine performance. However, above 80 m, the increase in initial investment due to additional tower height is not justified by the improvement in the capacity factor. As a result, the cost per kWh generated starts increasing beyond a tower height of 80 m.

The optimum tower height for a system is a site specific issue. Wind shear varies from place to place depending on the ground conditions. Apart from the shear, presence of trees or other obstructions in the wind flow path may demand taller towers.

There are some limitations for increasing the tower height. Taller towers are more visible and thus would aggravate the aesthetic issues of wind turbines. The tower height may also have to be restricted due to regulatory limits. The maximum permissible limit may vary from country to country. For example, in United States, this limit is 61.4 m. German regulations exempt a height upto 100 m from obstruction marking. If the turbines happen to be in prominent flight paths, the regulatory authorities may insist for navigation lights on taller towers. Unfortunately, any such markings may make the tower more visible to the public, thus causing visual annoyance. Further, extra heights added to the tower may cause its servicing and maintenance difficult, unless we have specialized lifts to reach the nacelle. Requirement of such devices may not be economically justifiable in many cases. Final decision on the tower height should be made after considering all these factors.

4.1.2 Rotor

Rotor is the most important and prominent part of a wind turbine. The rotor receives the kinetic energy from the wind stream and transforms it into mechanical shaft power. Components of a wind turbine rotor are blades, hub, shaft, bearings and other internals.

Blades of the wind turbine have airfoil sections as discussed in Chapter 2. Though it is possible to design the rotor with a single blade, balancing of such rotors would be a real engineering challenge. Rotors with single blade run faster and thus create undue vibration and noise. Further, such rotors are not visually acceptable. Two bladed rotors also suffer from these problems of balancing and visual acceptability. Hence, almost all commercial designs have three bladed rotors. Some of the small wind turbines, used for battery charging, have more number of blades- four, five or even six-as they are designed to be self starting even at low wind speeds.

Size of the rotor depends on the power rating of the turbine. The turbine cost, in terms of $ per rated kW, decreases with the increase in turbine size. Hence, MW sized designs are getting popular in the industry. For example, NEG Micon® has already installed a 4.5 MW prototype in Denmark [17]. The rotor of this turbine is built with massive 54 m blades. Some manufactures are planning even longer blades for their future projects. For example, the 61.5 m blade from the LM Glassfiber® is being installed in a 5 MW turbine with 125 m rotor, by the REpower Systems AG®, Germany.

Blades are fabricated with a variety of materials ranging from wood to carbon composites. Use of wood and metal are limited to small scale units. Most of the large scale commercial systems are made with multi layered fiberglass blades. Attempts are being made to improve the blade behavior by varying the matrix of materials, reinforcement structures, ply terminations and manufacturing methods. The traditional blade manufacturing method is open mold, wet lay-up. Some of the manufacturers are making their blades by vacuum assisted resin transfer molding (VARTM).

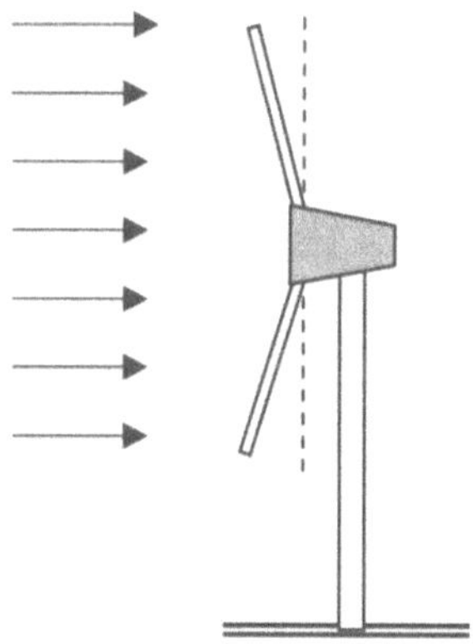

Fig. 4.7. Pre-bending of blades

With the increase in size, carbon-glass hybrid blades are being tried by some manufactures. These blades are expected to show better fatigue characteristics under severe and repetitive loading. The high stiffness characteristic of carbon reduces the possibility of blade bending in high winds and hence, they can be positioned close to the tower. Carbon also improves the edgewise fatigue resistance of the blades which is an advantage for bigger rotors. Weight of the blades can be reduced by 20 per cent by introducing carbon in the design. Usually, weight increases with the cube of the blade length. This can be restricted to an exponent of 2.35 in case of blades with carbon. A lighter blade demands lighter tower, hub and other supporting structures and thus may contribute to the economy of the system. Further, with the introduction of carbon in the design, we can twist couple the blades. Twist coupling improves the turbine performance by better power regulation and quick response to wind gusts. However, carbon-glass hybrid blades are costlier.

Most of the modern wind turbines have up-wind rotors. During wind loading, the blades of these rotors may be pushed towards the tower. This reduces the effective blade length and thus the rotor area. This back bending may also cause fatigue to the blades in due course. To avoid this problem, modern rotors have pre-bend geometries. The pre-bend blades stretch to its total lengths under wind loading, and thus exploit the full potential of the rotor as shown in Fig. 4.7.

We have discussed the basic aerodynamics of wind turbines in Chapter 2. The theories considered are the axial momentum theory, blade element theory and the strip theory. Based on these theories, a procedure was also evolved for the design of wind turbine blades. It should be noted that, we have discussed only the basic concepts of the design in these sections. Conditions under which a wind turbine is put to work are severe and unpredictable and thus cannot be fully defined by mathematical formulations. Further, the flow around the rotor is a very complex process, involving dynamic interactions between the fluid particles and blades elements. Expertise in atmospheric fluid dynamics, aerodynamics and structural dynamics is essential for understanding the behavior of the system under fluctuating conditions of wind regime. Hence, accurate and reliable prediction of wind rotor performance still remains a challenge to the scientists and engineers working in this area.

Various numerical methods are being followed by the researchers as well as industrial experts to design the rotor and estimate its performance. Some of the popular approaches are the Blade Element Momentum method [13,23], Vortex Lattice method [29] and Reynolds-averaged Navier stokes method [24,30]. Computer codes based on these models are also available for the design. All these methods have their inherent merits and demerits. For example, Blade Element Momentum and Vortex Lattice methods are strong in predicting the pre-stalled behavior of the rotor. However, they are inadequate in defining the stall and stall delay performances. Hence, these models may not be accurate in higher wind speeds. Reynolds-averaged Navier stokes method is preferred under these conditions. However, this model is inefficient in predicating the rotor vortex wake convection and boundary layer turbulence. Performances of these models under different flow conditions are compared in [7]. Further, based on the typical flow conditions, different variations of these models are also being used.

Pattern of rotor loading under fluctuating conditions and fatigue properties of materials used should also be considered in the design process. In wind farms, apart from the normal aerodynamic loads, interaction between the turbines may also cause unpredictable and excessive stress on the blades. The design process should consider the conditions of both extreme loading as well as repetitive loading. In case of extreme loading, the structural stability of the system against a single maximum expected load should be assessed. Usually, the sustainability of the system under an extreme wind load which may occur for 10 minutes, once in every 50 years, is considered.

In the repetitive load analysis, working of the system under a number of adverse conditions which are repeated during the course of operation should be analyzed. Frequency of occurrence of these conditions may vary and hence, they are weighed on the basis of respective occurrence probability [8]. In both the cases, the system response is assessed as a function of wind speed. There are certain areas of the blade which are more prone to fatigue and failure. For example, near the root section and one-third of the blade near the root are possible failure regions. Another problem is the buckling of the blade at the section of maximum chord, which is further propagated to cause blade failure.

The rotor loading may be analyzed either using parametric methods or empirical methods. In the parametric approach, the response of the systems to a given condition of loads is defined using the statistical models. As this method is based on sound statistical theories, it could be possible to extrapolate the system response under lower or higher frequency levels. Uncertainty levels of the loads can also be included in the analysis [16]. The Weibull distribution (explained in Chapter 3) is commonly used in the parametric approach.

Under the empirical approach, loads on the rotor under a set of simulated conditions are monitored. The system response is observed over a short period-usually 10 minutes-under predefined environmental conditions. This is then extrapolated to the full life time of the turbine. The major limitation of this method is that the validity of results is limited to the conditions under which the tests are conducted.

It may not be practical to include all possible combinations of operating conditions under the simulation process. Hence the parametric analysis is a better tool for estimating the fatigue loads of wind rotors.

Health monitoring of blades while in operation is a possible way to detect and rectify the blade fatigue. If the possibilities of failure are identified at an early stage, remedial measures can be taken, so that the failure zones are not combined and further propagated, causing the total blade collapse. Damage detection methods, like the stress wave monitoring technique, are being effectively used for the health monitoring of blades. A typical system consists of a set of piezo-ceramic patches generating stress waves across the possible failure regions, which are then received by sensors. The signals are further processed to detect the possibilities of fatigue at different blade sections.

The blades of the rotor are attached to hub assembly. The hub assembly consists of hub, bolts, blade bearings, pitch system and internals. Hub is one of the critical components of the rotor requiring high strength qualities. They are subjected to repetitive loading due to the bending moments of the blade root. Due to the typical shape of the hub and high loads expected, it is usually cast in special iron alloys like the spherical graphite (SG) cast iron. Forces acting on the hub make its design a complex process. Three dimensional Finite Element Analysis (FEA) and topological optimization techniques are being effectively used for the optimum design of the hub assembly [26].

The main shaft of the turbine passes through the main bearings. Roller bearings are commonly used for wind turbines. These bearings can tolerate slight errors in the alignment of the main shaft, thus eliminating the possibility of excessive edge loads. The bearings are lubricated with special quality grease which can withstand adverse climatic conditions. To prevent the risk of water and dirt getting into the bearing, they are sealed, sometimes with labyrinth packing. The main shaft is forged from hardened and tempered steel.

Replacing the bearings of an installed turbine is a very costly work. Hence, to ensure longer life and reliable performance, hybrid ceramic bearings are used with some recent designs. The advantages of ceramic hybrids are that they are stiffer, harder and corrosion free and can sustain adverse operating conditions. These bearings are light in weight and offer smoother performance. Electrically resistant nature of ceramic hybrids eliminates the possibilities of electrical arcing. These bearings are costlier than the standard bearings. However, they can be economical in the long run due to its better performance.

4.1.3 Gear box

Gear box is an important component in the power trains of a wind turbine. Speed of a typical wind turbine rotor may be 30 to 50 r/min whereas, the optimum speed of generator may be around 1000 to1500 r/min. Hence, gear trains are to be introduced in the transmission line to manipulate the speed according to the requirement of the generator. An ideal gear system should be designed to work smoothly

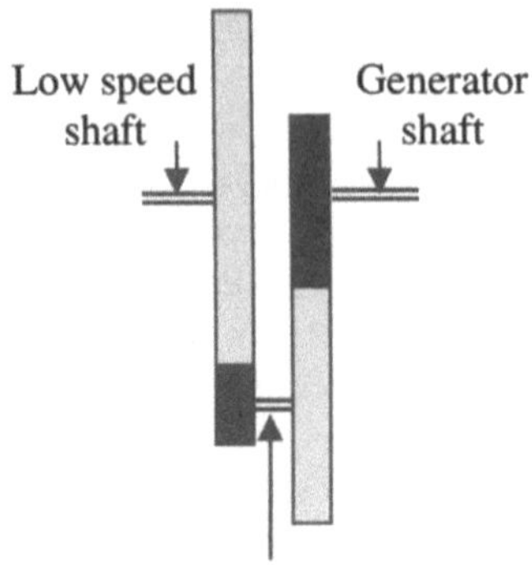

Fig. 4.8. Gear arrangements of a small wind turbine

and quietly-even under adverse climatic and loading conditions-throughout the life span of the turbine. Due to special constraints in the nacelle, the size is also a critical factor.

In smaller turbines, the desired speed ratio is achieved by introducing two or three staged gearing system. For example, the gear arrangement in a 150 kW turbine is shown in the Fig. 4.8. Here, the rotational speed of the rotor is around 40 r/min whereas the generator is designed to operate at 1000 r/min. Thus a gear ratio of 25 is required, which is accomplished in two stages as shown in the figure. Low speed shaft from the rotor carries a bigger gear which drives a smaller gear on the intermediate shaft. Teeth ratio between the bigger and smaller gear is 1:5 and thus the intermediate shaft rotates 5 times faster than the low speed shaft. The speed is further enhanced by introducing the next set of gear combination with bigger gear on the intermediate shaft driving a smaller gear on the generator shaft. Finally the desired 1:25 gear ratio is achieved.

If higher gear ratios are required, a further set of gears on another intermediate shaft can be introduced in the system. However, the ratio between a set of gears are normally restricted to 1:6. Hence, in bigger turbines, integrated gear boxes with a combination of planetary gears and normal gears are used. A typical gear box may have primary stage planetary gears combined with a secondary two staged spur gears to raise the speed to the desired level. By introducing the planet gears, the gear box size can be considerably reduced. Moreover, planet gears can reliably transfer heavy loads.

Due to its compact design, it is difficult to dissipate the heat generated during the power transmission in planetary gears. A typical 600 kW wind turbine generates waste heat equivalent to 18 kW during its full rate operation [25]. Another problem is the complexity in manufacturing these gears. Beveled gears are less noisy due to the smooth transfer of loads between the adjacent teeth. However, bevel gears cannot be used in planetary gear system.

Bearings for different points of the gear box are selected based on the nature of loads to be transmitted. For example, the planet carrier handles medium sized loads at lower speeds.

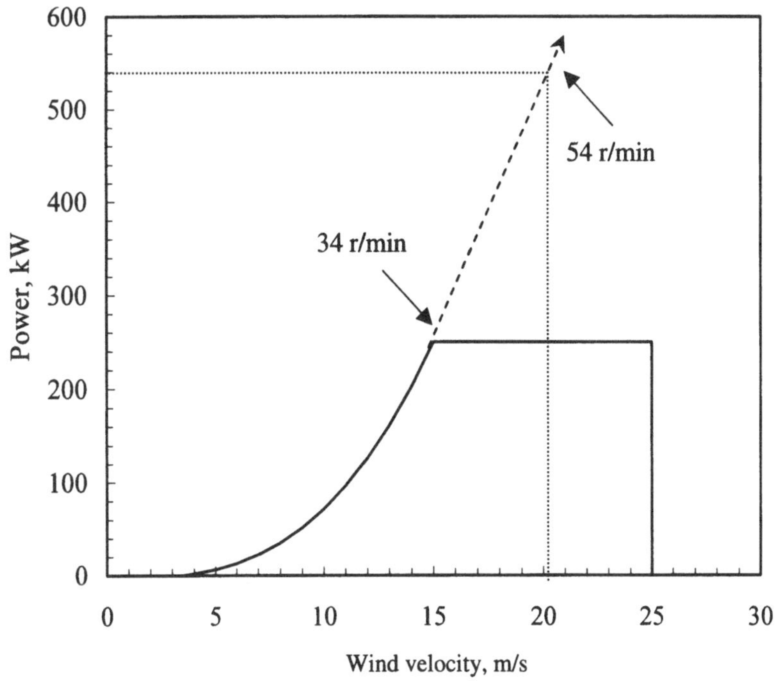

Fig. 4.9. Power curve of a typical wind turbine

Hence cylindrical roller bearings are recommended. The output shaft has to transmit high speed and low radial loads and hence cage guided cylindrical bearings may be used. Other possible options at this point are self aligning bearings or four point contact ball bearings.

Gears are designed on the basis of duration and distribution of loads on individual gear teeth. The load duration and distribution pattern (LDD), under a given set of wind conditions, are analysed. This is further extrapolated for the life time of the gears to arrive at the final design. Advanced numerical simulation tools are also being used for characterizing the dynamic response of wind turbine gears and other power trains [21,28].

4.1.4 Power regulation

Power curve of a typical wind turbine is shown in Fig. 4.9. The turbine starts generating power as the wind speeds crosses its cut-in velocity of 3.5 m/s. The power increases with the wind speed upto the rated wind velocity of 15 m/s, at which it generates its rated power of 250 kW. Between the rated velocity and cut-out velocity (25 m/s), the system generates the same rated power of 250 kW, irrespective of the increase in wind velocity. At wind velocities higher than the cut-off limit, the turbine is not allowed to produce any power due to safety reasons (refer Chapter 5 for more discussions on the power curve).

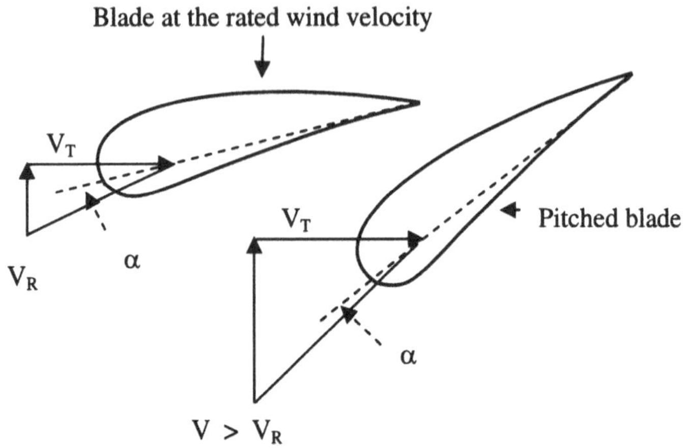

Fig. 4.10. Principle of pitch control

Power generated by the turbine is regulated to its rated level between the rated and cut-out wind speeds. If not regulated, the power would have been increased with wind speed as indicated by the dotted lines as in the figure. In the above example, we can see that the power corresponding to 20 m/s is more than twice the rated power of the system. However, if we want to harness the power at its full capacity even at this high velocity, the turbine has to be designed to accommodate higher levels of power. This means that, the system would require stronger transmission and bigger generator. On the other hand, probability for such high wind velocities is very low in most of the wind regimes. Hence, it is not logical to over design the system to accommodate the extra power available for a very short span of time.

Speed of the rotor also increases with the wind velocity. In the above example, the rotor speed increases from 34 r/min to 54 r/min, while the velocity changes from 15 m/s to 20 m/s. With further increase in velocity, the rotor may further speed up, finally reaching the run-away situation. It should also be noted that this increase in speed occurs in a short span of time, resulting in rapid acceleration. Hence it is vital that the power of the turbine should be regulated at constant level, at velocities higher than the rated wind speed. The common methods to regulate the power are pitch control, stall control, active stall control and yaw control.

As we have discussed in Chapter 2, wind turbine blades offer its maximum aerodynamic performance at a given angle of attack. The angle of attack of a given blade profile changes with the wind velocity and rotor speed. Principle of pitch control is illustrated in Fig. 4.10. Here V_R is the rated wind velocity, V_T is the velocity of the blades due to its rotation and α is the angle of attack. In a pitch controlled wind turbine, the electronic sensors constantly monitors the variations in power produced by the system. The output power is checked several times in a second. According to the variations in power output, the pitch control mechanism is activated to adjust the blade pitch at the desired angle as described below.

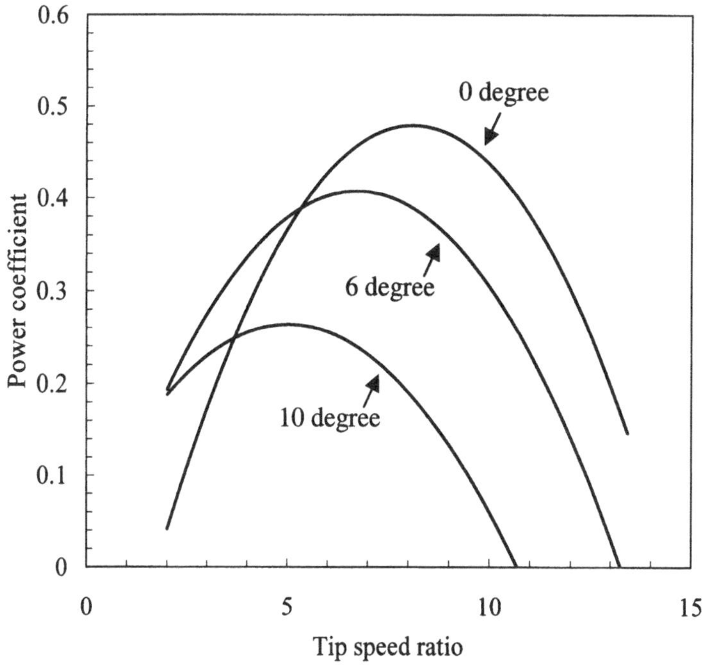

Fig. 4.11. Effect of blade pitch on the rotor performance

Between the cut-in and rated wind speeds, the turbine is made to operate at its maximum efficiency by adjusting the blade pitch to the optimum angle of attack. As the wind velocity exceeds V_R, the control mechanism change the blade pitch resulting in changes in the angle of attack as shown in the figure. From Fig. 4.11, we can see that, any changes in the angle of attack from its optimum level would in turn reduce the efficiency of the rotor. Thus, at wind speeds higher than V_R, we are shedding the excessive rotor power by spoiling the aerodynamic efficiency of the blades. Once the velocity comes down to the rated value or below, the blades are pitched back to its optimum position.

In a pitch controlled turbine, the blades are to be turned about their longitudinal axis by the pitch control mechanism in tune with the variations in wind speed. The pitch control mechanisms are driven by a combination of hydraulic and mechanical devices. Inorder to avoid sudden acceleration or deceleration of the rotor, the pitch control system should respond fast to the variations in wind velocity. Similarly, for maximum performance, the pitching should exactly be at the desired level. Thus, the pitch control system should be very sensitive.

Another method to regulate the power at high wind velocities is stall regulation. The basic principle of stall regulated turbines is illustrated in Fig. 4.12. In these turbines, profile of the blades is designed in such a way that when the wind velocity exceeds beyond the rated limit, the angle of attack increases as shown in the figure. With this increase in angle of attack, air flow on the upper side of the blade

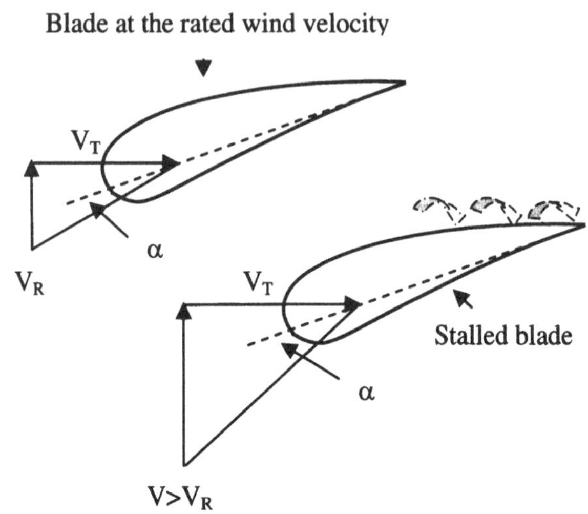

Fig. 4.12. Principle of stall control

ceases to stick on the blade. Instead, the flow starts whirling in an irregular vortex, causing turbulence. This kills the lift force on the blades, finally leading to blade stall. Thus, the excess power generated at high wind is regulated.

Pitch controlled turbines can capture the power more effectively in moderate winds as the blades can be set to its optimum angle of attack by pitching. However, moving components are to be introduced in the blade itself for adjusting its angle, which is a drawback of these systems. Similarly, the control unit should have high sensitivity towards wind fluctuations which makes them costlier.

On the other hand, stall controlled blades do not require any control system or pitching mechanism. However, the blades are to be aerodynamically twisted along its longitudinal axis. Design and manufacturing of such blades demand sophistication. Structural dynamics of the system should be carefully analyzed before the design to avoid any possible problems like the stall induced vibrations. Power curve of a typical stall controlled turbine is shown in Fig. 4.13. Performance of these turbines at higher wind speeds is not impressive as the power falls below the rated level. In spite of these limitations, many wind power plants are still installed with stall controlled turbines.

Some modern turbines exploit the advantages of both the pitch and stall controlled options for regulating its power. This is called the active stall controlled power regulation. In this method, the blades are pitched to attain its best performance in lower winds. However, once the wind exceeds the rated velocity, the blades are turned in the opposite direction to increase the angle of attack and thus forcing the blades into a stall region. The active stall allows more effective power control and the turbine can be run nearly at its rated capacity at high winds.

Another method of power control is to push the rotor partly away from the wind direction at higher wind speeds. This is called the furl or yaw control. The rotor spin axis is pushed to an angle to the incoming wind direction at high winds for regulating the power. When the velocity exceeds the cut-off limit the axis

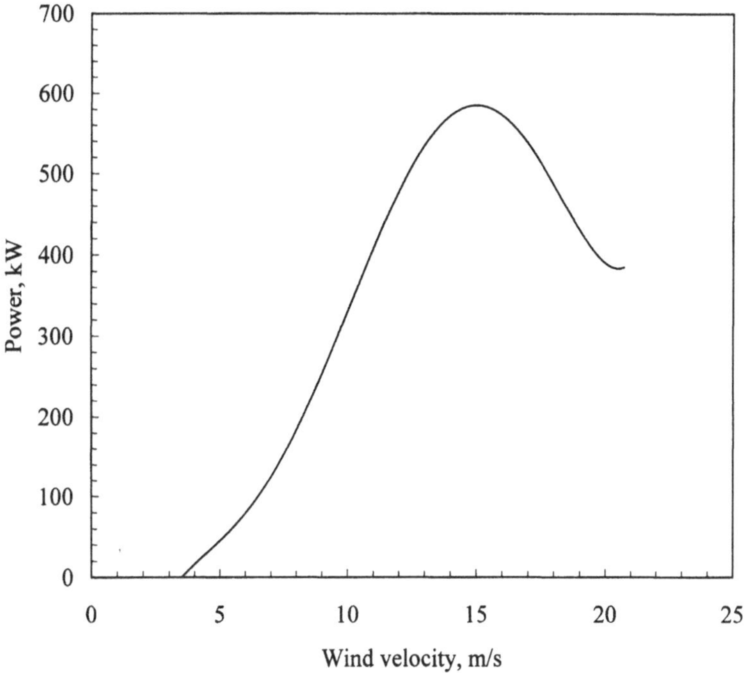

Fig. 4.13. Power curve of a typical stall controlled wind turbine

position would be nearly perpendicular to the wind vector, thus ceasing the power generation. When the rotor is partially yawed, it experiences cyclic stresses. Due to this reason, yaw control is employed only for small wind turbines.

4.1.5 Safety brakes

During the periods of extremely high winds, wind turbines should be completely stopped for its safety. Similarly, if the power line fails or the generator is disconnected due to some reason or the other, the wind turbine would rapidly accelerate. This leads the turbine to run-away condition within a few seconds. For example, consider the turbine in Fig. 4.14. Rated operating speed of this turbine is 34 r/min. Under run-away condition, the rotor speed of this turbine shoots upto 90 r/min within five seconds. The system is not designed to tolerate such a high speed and resulting acceleration. Hence, the turbine should essentially be fitted with safety devices, which will break the system and bring it to halt under such conditions.

As the rotor accelerates rapidly, the safety brakes should have rapid reactive response to prevent the run-away condition. Two types of brakes are commonly used with wind turbines. They are aerodynamic brakes and mechanical brakes. Inorder to ensure the safety, wind turbines usually have two braking systems, one functioning as the primary brake and the other as a backup option which comes in to action if the primary system fails.

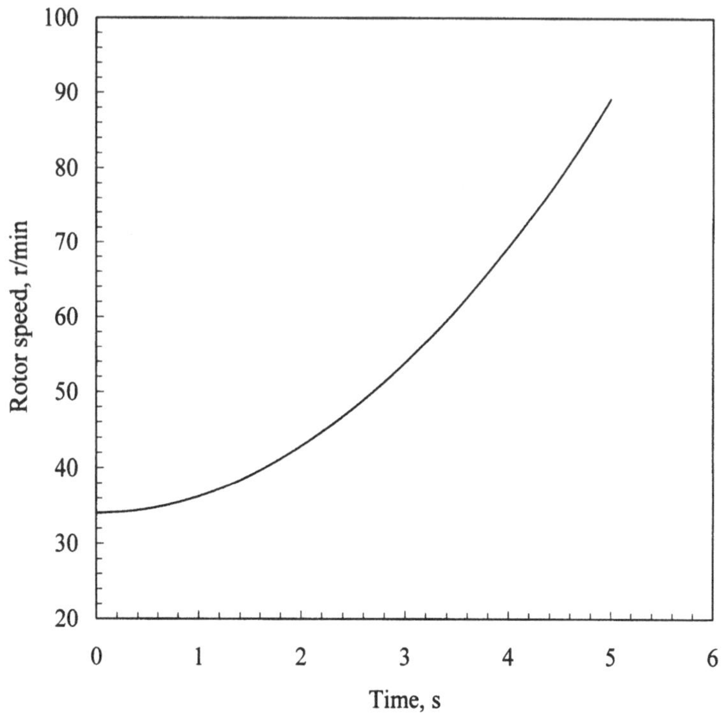

Fig. 4.14. Acceleration of the rotor at run-away condition

Aerodynamic brakes are the primary system in most of the wind turbines. Aerodynamic braking in pitch and stall controlled turbines are different. In pitch and active stall controlled systems, the entire blade is turned 90° along its longitudinal axis, there by hindering the driving lift force. Thus the rotor would stop after making a few more rotations.

In contrast, it is the tip of the blade which is moved in stall controlled turbines. Position of the blade tip, relative to the blade, can be changed using a shaft and bearing assembly fixed inside the main body of the blades. During normal operation, the tip is held in position with the blade using hydraulic force. When the blades are to be stopped, the hydraulic force that keeps the tip in line with the blade is cut-off, there by allowing the blade tip to move outwards. Driving unit of the blade shaft is then activated which turns the tip through 90°. This brings the blades to the braking position. Although the blades are not completely stopped by tip braking, the rotor can be brought to a freewheeling speed, which is much lower than its normal operating speed. Once the dangerous situation is over, the blades are brought back to the working position by the hydraulic system. Field experience shows that the aerodynamic braking is quite effective in protecting the turbines.

In addition to the aerodynamic braking, a mechanical brake is also provided with the turbine as a back up system. These brakes are applied to bring the rotor to 'full stop' position in stall controlled turbines. They are also useful to lock the rotor during the turbine maintenance.

Mechanical brakes are friction devices, consisting of brake disc, brake caliper, brake blocks, spring loaded activator and hydraulic control. The brake disc is fixed to the high speed shaft coming from the gear box. Under normal operation, the brake disc and blocks are held apart by hydraulic pressure. When the system is to be braked due to safety reasons, this pressure is released and the brake spring presses the block against the disc. This will bring the system to halt. Being frictional devices operating under extreme loading, the brake blocks are made with special alloys which can tolerate high stress and temperature.

Possibilities of introducing electro-dynamic braking system are also being investigated in large scale fixed speed wind turbines. These brakes, with capacitors and resistors connected to the terminals of rotating induction machine, work on the principle of self excitation. Required braking torque is developed by the current and voltage induced in the machine windings. The braking torque, being a function of rotational speed, can effectively be used for protecting the turbine from reaching the run-away condition [9].

4.1.6 Generator

Generator is one of the most important components of a wind energy conversion system. In contrast with the generators used in other conventional energy options, generator of a wind turbine has to work under fluctuating power levels, in tune with the variations in wind velocity. Different types of generators are being used with wind machines. Small wind turbines are equipped with DC generators of a few Watts to kilo Watts in capacity. Bigger systems use single or three phase AC generators. As large-scale wind generation plants are generally integrated with the grid, three phase AC generators are the right option for turbines installed at such plants. These generators can either be induction (asynchronous) generators or synchronous generators.

Induction generator

Most of the wind turbines are equipped with induction generators. They are simple and rugged in construction and offer impressive efficiency under varying operating conditions. Induction machines are relatively inexpensive and require minimum maintenance and care. Characteristics of these generators like the over speed capability make them suitable for wind turbine application. As the rotor speed of these generators is not synchronized, they are also called asynchronous generators. Induction machines can operate both in motor and generator modes. Let us first consider it as a motor and then look into the generator option.

The cross sectional view of an induction motor is shown in Fig. 4.15. Stator consists of a number of wound coils placed inside its slots. The stator windings are connected to the power supply. They are wound for a specified number of poles depending on the speed requirement.

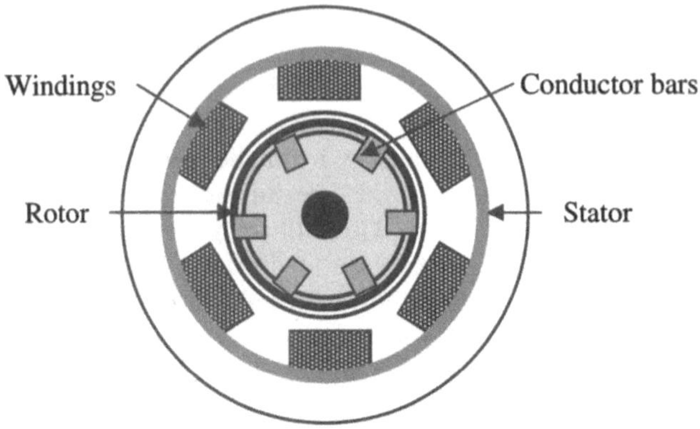

Fig. 4.15. Cross-sectional view of an induction motor

The motor speed decreases with the increase in number of poles. The rotor of the machine has a cylindrical laminated core with slots for housing rotor conductors. The conductors, which are aluminum or copper bars, are individually placed in the slots. The bars are electrically connected with thick aluminum end rings, thus making them permanently short circuited. As this construction resembles to a squirrel cage, they are often named as squirrel cage rotors. Induction motors with wound rotors are also available.

When the stator windings are fed with three phase supply, a magnetic flux, rotating at uniform speed (synchronous speed), is developed. The flux passes through the air gap and reaches the 'now' stationary rotor and cuts its conductors. Due to the relative speed between the rotating flux and the stationary conductors, a strong electro-magnetic force is induced in the rotor. Magnitude of this induced electro-magnetic force is proportional to the relative velocity between the flux and the rotor. Thus, a very strong current is induced in the rotor bars as they are short circuited by the end rings. This current, according to Lenz's law, would tend to oppose the very cause of it - that is the relative velocity between the flux and the rotor. Hence, to reduce the relative velocity, the rotor starts rotating in the same direction of the flux.

Although the rotor tends to chase the rotating flux to reduce the relative velocity, it never succeeds in attaining the same speed due to frictional losses (if it succeeds, there will not be any relative speed between the two and hence no induced electro-magnetic force or current). Hence, the rotor always rotates at a speed slightly lower than the synchronous speed. The difference between the synchronous speed (N_s) and the rotor speed (N_R) is termed as the slip of the motor. Thus, the slip (S) is given by

$$S = \frac{N_s - N_R}{N_s} \tag{4.3}$$

The synchronous speed of the induction motor is given by

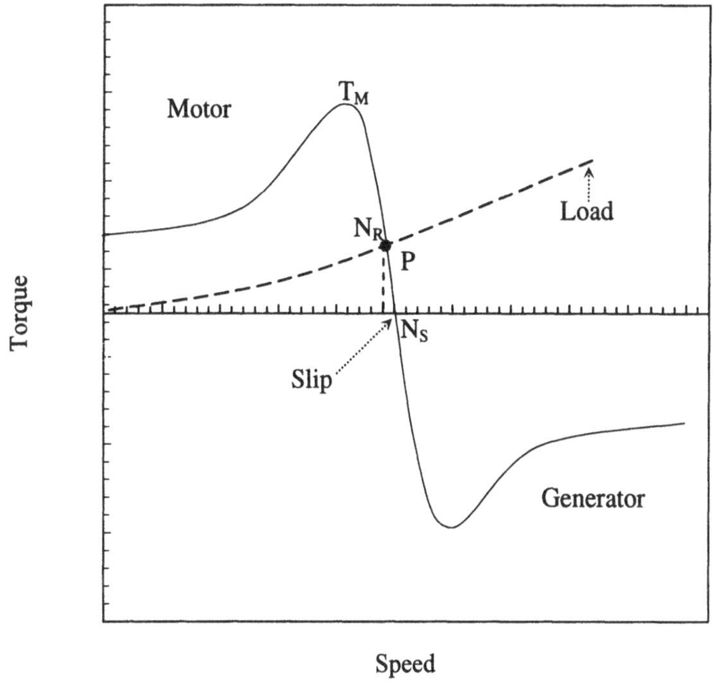

Fig. 4.16. Characteristics of a typical induction machine

$$N_s = \frac{120\,f}{P} \tag{4.4}$$

where f is the frequency and P is the number of poles. Thus, an induction motor with 4 poles, connected to a grid of 50 Hz current, will have a synchronous speed of 1500 r/min. With 1 per cent slip, the speed of the rotor is 1485 r/min. From the expression given above, we can see that the speed of the motor can be manipulated by changing the number of poles. In the above example, if we use a six pole motor, then the synchronous speed can be reduced to 1000 r/min.

Now let us examine the operation of an induction machine as the generator. Typical torque speed curve of the machine is shown in Fig. 4.16. The dotted line shows the variation in torque demand of the connected load. At start (N=0), the torque of the load is less than that is developed by the motor, causing it to accelerate. With increase in speed, the motor torque increases upto its maximum level T_M, beyond which the torque rapidly drops.

The torque-speed curves of the motor and the load are intersecting at a point P. This is the steady state condition for the given load and the system will settle to work at this point. N_R is the speed corresponding to P, which is less than the synchronous speed N_S. The difference between N_S and N_R is the slip. At this phase of operation, the machine acts as a motor.

When we couple this machine with a grid integrated wind turbine, initially it draws current from the grid as in case of a motor. The speed picks up and the rotation of the wind turbine causes the system to exceed the synchronous limit N_S. Thus, rotor moves faster than the rotating magnetic field. At speeds higher than N_S, the torque is negative as seen from the figure. Thus, current flows in the opposite direction, that is from the system to the grid. Thus the machine functions as a generator when it is driven by an external prime mover, like the wind turbine in our case.

As the working speed exceeds the synchronous speed in the generator phase, the slip is negative. Generally the slip is in the range of 1-3 per cent. Thus, a four pole machine with synchronous speed 1500 r/min and 50 Hz frequency would turn at 1515 r/min (for 1 per cent slip), while functioning as the generator. The generator speed depends on the torque applied by the turbine. With fluctuations in torque, the generator speed is adjusted slightly. This is a highly desirable property for a generator, coupled with a fluctuating source like wind. It helps in minimizing the load imposed on the transmission components.

External excitation is essential for an induction generator before it is put to work. For grid integrated systems, this is not a limitation as it can draw the required current from the grid. However, in stand alone mode, external devices like capacitors or batteries are required to provide the necessary excitation current to the generator.

During calm periods, the wind turbine is idle and there is no external driving force for the generator. Hence, the system will draw current from the grid and start behaving as a motor, driving the rotor like a huge fan. To avoid this situation, the generator is disconnected from the grid during calm periods. At wind speeds lower than the cut-in speed, the rotor is allowed to rotate freely without generating any power. At wind speeds preset on the basis of the cut-in velocity, the system is gradually connected to the grid to receive excitation and then to function as a generator as explained above. This is accomplished by special electronic contacts called thyristors. Thus, thyristors allow gradual loading of the turbine and thus eliminate the possibilities of transmission failure and grid damage due to sudden surging of current. Once the system is cut-in, the thyristors are bypassed using bypass switches. Then the flow will be directly from the generator to the grid, avoiding power loss due to the thyristors.

Synchronous generator

As we have seen, speed of the rotor lags behind the synchronous speed in an induction generator. In contrast, rotor and magnetic field rotates at the same speed in a synchronous generator. Cross-sectional view of a synchronous generator, in its simplest form, is shown in Fig. 4.17. It consists of a rotor and a three-phase stator similar to an induction generator. The stator and rotor have the same number of poles. The generator shown in the figure has two poles. The stator has coils wound around them, which are accommodated in slots as shown in the figure. The stator windings are displaced circumferentially at 120° interval.

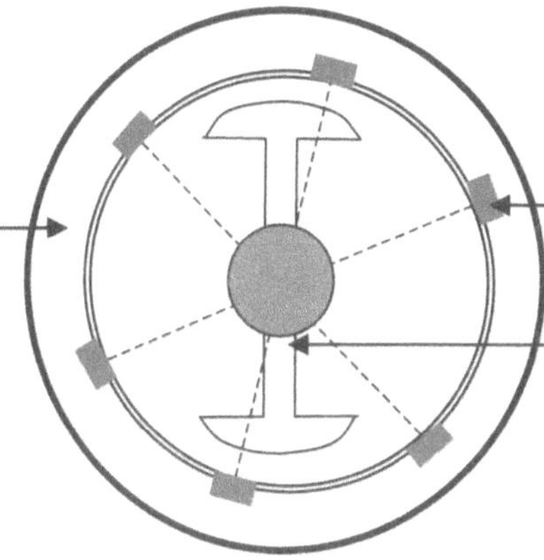

Fig. 4.17. Principle of synchronous generator

Rotors with either permanent magnet or electromagnet can be used in this generator. Permanent magnets are costly and they loose their magnetic properties while working under powerful magnetic fields. For this reason, most of the synchronous generators coupled with wind turbines are employed with electromagnets. These electromagnets are magnetized by feeding direct current to the rotor field windings. Brush type or brushless exciters may be used to magnetize the rotor. The magnetization current may be drawn from the grid also. As the grid supplies current in alternating form, it is converted to DC before using it for magnetization. This excitation establishes strong north-south magnetic characteristics in the rotor poles.

As the wind turbine turns the rotor, the magnetic field linking the stator coil varies sinusoidal with time, in proportion with the rotor speed. The rate of change of flux induces a sinusoidal voltage in the coils. As the coils are displaced at 120°, a balanced three-phase supply is generated, which is fed to the grid.

As seen from Eq. (4.4), the speed of the synchronous generator depends on the grid frequency and number of poles. For example, in a 50 Hz grid, above generator with 2 poles will run at

$$\frac{120 \times 50}{2} = 3000 \text{ r/min}$$

A generator with 6 poles would run at 1000 r/min whereas with 24 poles, the speed can be reduced to 250 r/min, and so on. Generators with six poles are commonly used with wind turbines.

For wind energy generation, low speed generators are advantageous as the gearing required for the system can be minimized. Taking the average rotational speed of a wind turbine as 40 r/min, can we eliminate the gear box by designing a generator with 150 poles? Theoretically it may sound possible, but the practical viability of gearless generators is yet to be established. As we introduce more poles in the generator, it becomes bulky and expensive.

4.1.7 Fixed and variable speed operations

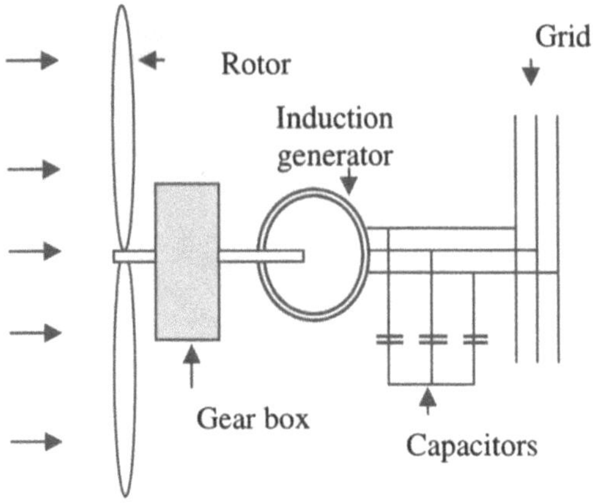

Fig. 4.18. Fixed speed wind turbine

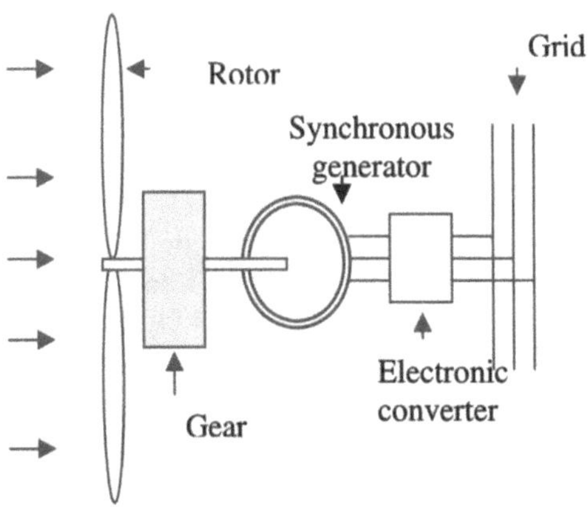

Fig. 4.19. Variable speed wind turbine with synchronous generator

A wind turbine can be designed to run at fixed or variable speeds. In fixed speed turbines, the rotor is coupled with an induction generator via speed increasing gears as shown in Fig. 4.18. Stator winding of the generator is directly wired to the grid. As we have seen, induction generators require excitation power from the grid. This may result in undesirable voltage variations, especially in weaker networks. To avoid this problem, capacitors are provided in the circuit as shown in the figure. With this configuration, the wind turbine will run at constant (or nearly constant) speed and feed the grid with power in a predetermined frequency (50 Hz or 60 Hz).

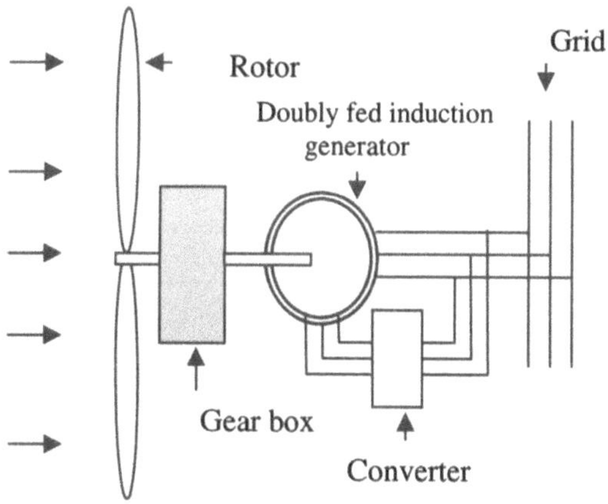

Fig. 4.20. Variable speed wind turbine with doubly fed induction generator

Wind turbines which operate in variable speeds are equipped with either synchronous generators or a doubly fed induction generators. In systems with synchronous generator, the operating speed changes randomly with fluctuations in the wind velocity and hence the output voltage and frequency would also vary. This output cannot be directly fed to the grid due to its poor power quality. Thus the wind turbine is totally decoupled from the grid in the variable speed option. The power is fed to the grid after conditioning through a suitable interface (Fig. 4.19). Thus, the AC generated by the synchronous generator is first rectified into direct current and then inverted back to AC at standard grid frequencies (50 Hz or 60 Hz), before feeding it to the grid.

In variable speed turbines with doubly fed induction generators, the stator winding is directly connected to the grid. However, the rotor winding is fed through a converter which can vary the electrical frequency as desired by the grid. Thus the electrical frequency is differentiated from the mechanical frequency, which allows the variable speed operation possible. Typical configuration of such a system is shown in Fig. 4.20.

The fixed speed turbines are simple in construction and thus tend to be cheaper than the variable speed option. However, as it cannot track the ever fluctuating wind speed, the energy capture is not as efficient as in fixed speed systems. As we have discussed, a wind rotor offers maximum power coefficient at its design tip speed ratio. For constant speed operation, this maximum power coefficient can only be attained at the design wind velocity of the turbine. This is illustrated in Fig. 4.21. Here, V_1 is the design wind velocity at which the turbine operates at a speed of N_1 and generates a power of power P_1. At this wind speed, the turbine works at its maximum power coefficient. However, when the wind speed varies from V_1 to say V_2, as the fixed speed system could operate only at the same speed N_1, the power developed is P_2 as in the figure. Thus, the peak power of P_3 developed by the rotor at V_2 is not fully utilized by the system as the turbine has to run at its fixed speed N_1.

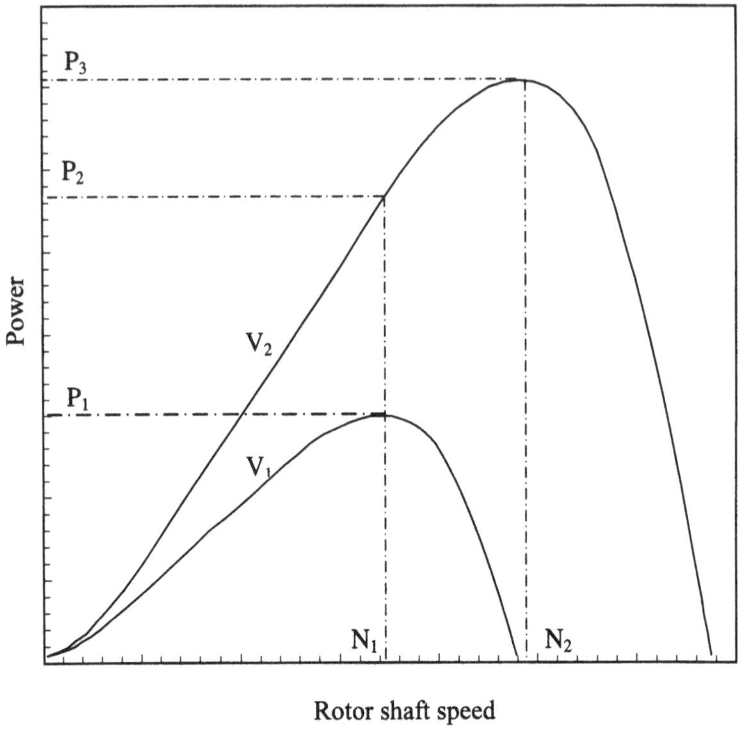

Fig. 4.21. Rotor speed versus power at two different wind velocities

On the other hand, if variable speed operation is possible, the system can be designed to run at a speed of N_2 at velocity V_2, and thus generating the highest possible power P_3 at this velocity. Thus the variable speed option allows the turbine to operate at its optimum level at a wide range of wind speeds. In other words, by allowing the turbine to operate at variable speed, the energy capture can be maximized. This is highly beneficial at locations with highly fluctuating wind regimes. The energy capture can be 8 to 15 per cent more in some specific sites.

Variable speed operation also reduces the load on the drive trains. Operating the turbine at constant speed under varying power levels results fluctuations in the torque transmitted. On the other hand, if speed is allowed to vary, the torque can be held almost constant over a wide range of power. This alleviates the stresses imposed on the structural components and thus, the turbine can be made lighter and cheaper. The rotor operating at variable speed can absorb the excess energy due to wind gust by speeding up and give it back to the system while the wind slows down. They offer better power quality as the flicker problem is minimized. As the rotational speed is low at lower wind speeds, variable speed turbines are less noisy making them acoustically acceptable.

However, variable speed turbines require complex and costly power electronics. Another problem is the excitation of structural resonance. In spite of these, variable speed design is getting popular these days. Several commercial manufactures are shifting to the variable speed option, especially for their offshore designs.

4.1.8 Grid integration

Till the recent years, contribution of wind power to the grid was not significant to affect the power quality. Hence, integrating wind energy to the power grid was not a matter of concern. Share of wind generated electricity in the grid is increasing day by day. For example, wind energy penetration in some areas of Europe has reached to a level of 10 to 15 per cent. In some specific pockets in Denmark, it is as high as 80 per cent [11]. Grid penetration at this level definitely warrants careful integration of wind energy to the electrical network.

The quality of the electrical network to which wind energy is fed is one of the important factors to be considered in the grid integration. The output frequency has to be maintained very closely to 50 or 60 Hz, depending on the local needs. The grid should be strong enough to withstand the characteristics of wind generated electricity. Instability in the grid can cause the wind farm to shut down. This becomes more critical if the penetration rate is high.

Matching the supply with the demand is quite important. Usually the supply system may consist of a number of power plants connected to the network. This may include nuclear, combined cycle gas turbine, hydro-electric plants, coal etc, along with wind turbines. Some of these plants provide the base load where as others may be utilized as the load tracking plants. Due to obvious economic reasons, the plants with high initial cost and low running cost are chosen to meet the base load where as systems requiring lower capital and higher running cost meets the additional demand.

For example, the load on a grid, at different time scales of a typical day, is shown in Fig. 4.22. The demand varies significantly with time, with its peak at certain hours of the day. Here, nuclear and combine cycle gas turbine, along with wind, supplies the base load. The coal based plant is regulated to meet the excess load.

Wind, like other renewables, is an ideal base load supplier. Apart from economic criteria, environmental issues and regulatory requirements are also in favor of using wind generated electricity to meet the base load. As the running cost of wind turbines is negligible in comparison with technologies like coal, it is advantageous to utilize the turbines at their full potential throughout the day. In many parts of the world, it is obligatory for the utility company to purchase all the electricity generated from wind. In countries like UK, wind energy has the ‘must take’ privilege under the Non Fossil Fuel Obligation, which demands the utilities to meet certain percentage of their electricity supplies from renewables.

For regulating the generation, demand at different time’s scales has to be accurately forecasted. There are short time variations in demand ranging from several seconds to 10 minutes as well as long time variations involving 10 minutes to several hours. Short term variations are often regulated by adjusting the generators. Variations in load are sensed and signals are sent to the control system to adjust the generators output. Momentary increase in load can also be adjusted by allowing the voltage to drop slightly, say up to 3 per cent. Similarly, the sudden drop in demand can be adjusted by allowing slight increase in the voltage.

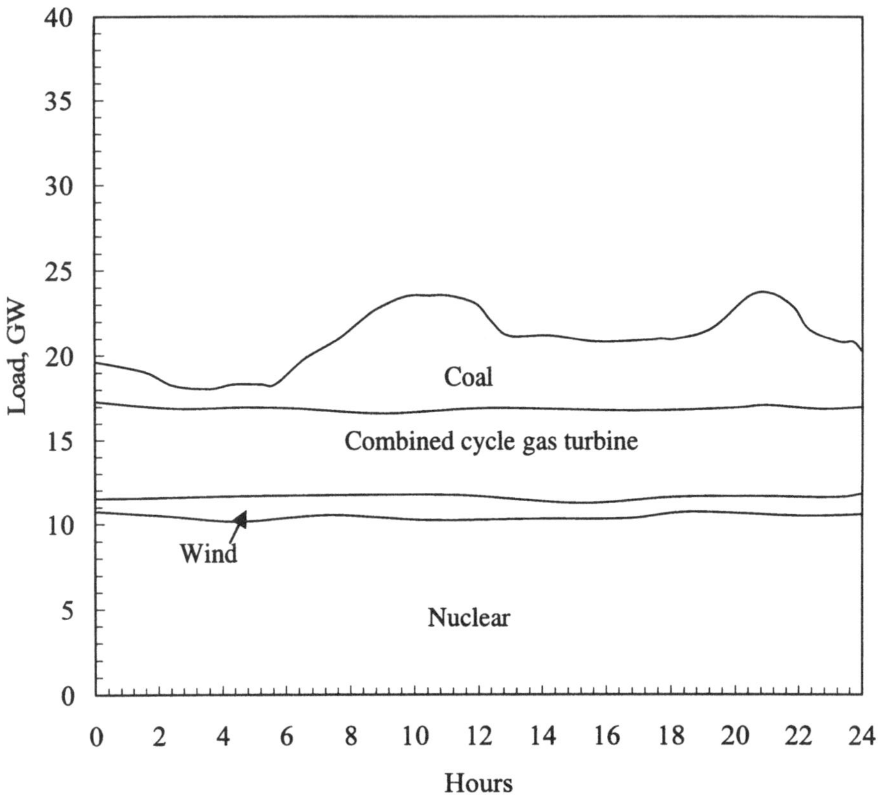

Fig. 4.22. Typical daily variation in the electric demand and generation

However, for meeting the long term fluctuations, an economic decision is taken on the generating option in response to the demand side variations. The demand may be peak at certain hours of the day-for example 8 AM to mid day as in the figure. Utility operator should be ready to provide this excess demand by increasing the share of load tracking source (coal in the above example).

A certain time period is required for starting the generators and synchronizes it with the grid. Hence the fluctuations in the load are to be predicted beforehand. This will help us to decide which system should be committed to the grid, at what time. Demand prediction models are used for this purpose. These models use the historical demand data along with meteorological predictions (to forecast chills etc.) and daily event schedules (holidays/working days or any special programmes in the TV etc.) to arrive at load estimations. Accuracy of these forecasting tools is found to be reasonably acceptable. Side by side, the power available from the wind turbines integrated to the grid also has to be estimated to ensure the share of wind generated electricity in the supply system. Availability of useful wind at the turbine site is estimated using meteorological models. HILRAM model of the Danish Meteorological Institute is an example. This information is fed to wind energy models (RISØ system or IMM system etc.) and the results are delivered to the utilities [22]. This process is schematically presented in Fig. 4.23.

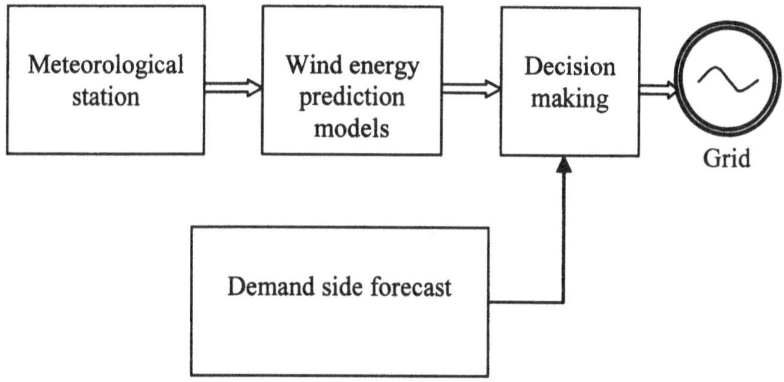

Fig. 4.23. Process of integrating wind power to the grid

4.2 Wind farms

Wind turbines of various sizes are available commercially. Small machines are often used for stand alone applications like domestic or small scale industrial needs. When we have to generate large quantities of power, several wind turbines are clubbed together and installed in clusters, forming a wind farm or wind park. There are several advantages in clustering wind machines. The installation, operation and maintenance of such plants are easier than managing several scattered units, delivering the same power. Moreover, the power transmission can be more efficient as the electricity may be transformed to a higher voltage.

Several steps are involved in the successful planning and development of a wind farm. They are:

1. preliminary site identification
2. detailed technical and economical analysis
3. environment, social and legal appraisal and
4. micro-siting and construction

The first step in the development of a wind farm is to identify a suitable location, having reasonably high wind velocity. Once the broad geographical region for the development of the proposed wind farm is identified, it may be possible to locate several sites which could be used for constructing the wind farm. Wind data available from local weather stations, airports etc. or published documents like wind maps may be used for this purpose. A candidate site must usually have a minimum annual average wind speed of 5 m/s. Once such sites available in the region are identified, computer models are used for estimating the energy potential of these sites in different time frames. The WERA model, provided with this book, can be used for this purpose. Thus the candidate sites are compared and short listed on the basis of their wind potential.

Fig. 4.24. A land based wind farm (Courtesy of Renewable Energy Systems Ltd., www.res-ltd.com)

In the first stage we relied on existing information to rate the potentiality of the sites. In the next stage, more rigorous analysis is required. The nature of the wind spectra available at the sites is to be thoroughly understood for the detailed technical analysis. For this, wind speed has to be measured at the hub height of the proposed turbines. Anemometers installed on guyed masts are used for wind measurement. Details of anemometers and measurement techniques are described in Chapter 3. Several anemometers, installed at different locations, are preferred for accurate analysis. Wind speed at the site has to be monitored at least for six months. If time and resources permit, the duration may further be increased to one year or even more.

Once long term wind data collected over short periods are available, detailed wind resource analysis is possible using computer programmes like WERA. As discussed in the next chapter, the total kWh output from a turbine, at a given site, depends on the matching between the site's wind profile and the machine characteristics. Hence, it is advisable to include the behavior of the turbines, proposed to be installed at the site, in the performance model. This gives us the capacity factor and kWh output from the turbines over different time intervals.

Apart from the sites wind potential, other factors like access to the grid, roads and highways, existing infrastructure for power transmission and ground condition at the site are also to be critically analyzed while choosing the site. The local electricity distribution system at these sites should be examined to ensure that minimum infrastructures are additionally required for feeding the power to the grid.

Similarly, accessibility to major highways and roads is also an important factor, as we have to transport the turbine and its components to the site. Availability and cost of land for the wind farm development is another major consideration. If our intention is to sell the generated power, an understanding on the prevailing energy market is also essential. The physical condition of the site should be thoroughly examined at this stage. This will give us an idea about the cost of foundation and other related constructions. The size (power rating) and number of turbines required for the project can be decided. Cost of the turbines and its accessories as well as the mode of maintenance may be negotiated with the manufacturer or local suppliers. Once all these issues are examined and costs involved are estimated, we can further workout the economics of wind energy generation at the sites as discussed in Chapter 7.

For bringing the project to a reality, apart from being economically viable, it should be environmentally acceptable. The major concerns are visual effects, avian interaction, noise emission and ecological factors. Detailed discussions on these aspects are presented in Chapter 6. Local survey and consultation with the local planning authority would be helpful in determining the environmental acceptability of the project. It should also be ensured that the proposed project is acceptable to the local residents. The developer should discuss the proposed project with the local community for avoiding any possible hassles in a later stage. It should be ensured that the project comply with the statutory requirements prevailing in the region. The sites, which do not meet the requirements in the above aspects, may be dropped from the 'short list'.

Finally, the site satisfying all these requirements-technical, economical, environmental, legal and social-in the best possible way can be selected for the wind farm development. Next step is to formulate a detailed proposal for the project. Suitability of the site in all the above aspects should be described in the proposal. The proposal may then be submitted to the appropriate body (for example, the local planning authority) for further processing and approval. As development of wind farm is a capital intensive affair (approximately costing $ 1 million per MW of installed capacity), possible funding for the project also should be located side by side.

Once the proposal for the project is approved by the competent authority, then we can proceed further with the micro sitting. Micro siting involves laying out the turbine and its accessories at optimum locations at the selected site. Turbines are placed in rows with the direction of incoming wind perpendicular to it, as shown in Fig. 4.25.

When several turbines are installed in clusters, the turbulence due to the rotation of blades of one turbine may affect the nearby turbines. Inorder to minimize the effect of this rotor induced turbulence, a spacing of 3 D_T to 4 D_T is provided within the rows, where D_T is the rotor diameter. Similarly, the spacing between the rows may be around 10 D_T, so that the wind stream passing through one turbine is restored before it interacts with the next turbine. These spacing may be further increased for better performance, but may be expensive as we require more land and other resources for farther spacing.

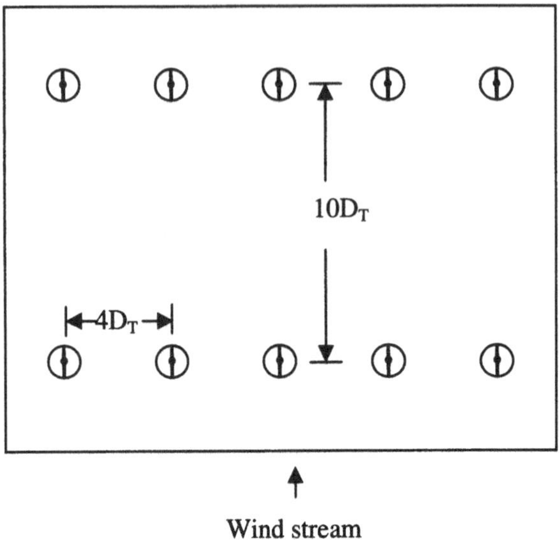

Fig. 4.25. Typical layout of a wind farm

It is a usual practice to leave a clearance of ($h_T + D_T$) from the roads, where h_T is the hub height of the turbine. Leaving this clearance on both sides, the row width available at a given site can be calculated. If not constrained by other factors, the number of turbines per row (N_{TR}) may be estimated as

$$N_{TR} = \frac{L_R}{S_R} + 1 \tag{4.5}$$

where L_R is the row length and S_R is the row spacing. If P_F is the total capacity of the wind farm and P_T is the rated power of a turbine, then

$$N_T = \frac{P_F}{P_T} \tag{4.6}$$

where N_T is the total number of turbines in the farm. Hence the total number of rows is obviously

$$N_R = \frac{N_T}{N_{TR}} \tag{4.7}$$

Although the above calculations can give us some indication on the turbine placement, in practice, the final placement of individual turbines at a given site depends on several factors like the shape and size of the available land, existing electrical network etc. Turbines may be relocated due to environmental factors also-for example we may violate the above norms to locate turbine at a point which is less prominent from a visually critical spot. British Wind Energy Association has formulated a comprehensive guideline for wind farm development [5].

4.3 Offshore wind farms

Fig. 4.26. Middelgrunden offshore wind farm, Denmark (Courtesy of Siemens Wind Power A/S, www.siemens.com/powergeneration)

In the recent years, there is a strong trend to shift the wind farm activities to offshore. Offshore wind farms are not a new concept as the possibilities for such projects were explored even during the 1970s and 1980s. Though these studies could bring out many critical issues related to the offshore wind potential and its exploitation, the real offshore hardware was realized with the commissioning of the Helgoland project in Germany during 1989. This was followed by several projects like Blekinge, Sweden (1990) and Vindeby, Denmark (1991). Today, a number of ambitious offshore projects are in the pipeline. At the current growth rate, worldwide contribution from offshore projects is estimated to be 4,500 MW by 2010 [6]

Offshore wind energy projects have several attractions. Offshore winds are stronger and steadier than the onshore wind. For example, at 10-15 m from the shore, the velocity of wind is higher by 20-25 per cent. As the wind velocity at the site critically influences the energy generated and thus the cost of unit generation, this stronger wind spectra is a significant advantage. Further, in comparison with the onshore wind, offshore wind is less turbulent. This reduces the fatigue loads on turbines and thus increases their service life. As we can see in Chapter 7, an increase in project life means reduction of the generation cost. Similarly, the sea offers lesser resistance to the wind flow, resulting in lower wind shear. This means that taller towers may not be required for the offshore projects (However, there may still be an increase in wind speed with height at offshore, so an economically optimum tower height is recommended).

Offshore wind projects are more environmentally acceptable as the land use, noise effect and visual impact may not be major concerns for planning approvals. Land based turbines are often made to run at tip speed ratios lowers than the optimum to reduce the noise pollution. Offshore systems can be designed to operate at higher speeds-some times 10 per cent higher-resulting in better aerodynamic efficiency. Similarly, offshore units are bigger in size and hence have better economics. Currently, bigger turbines of 2 MW or more capacity are being used for offshore installations. As aesthetic is not a major concern at offshore, it is possible to go for two bladed rotors instead of the conventional three bladed designs. This not only reduces the weight, but also improves the aerodynamic efficiency.

For offshore installation, turbines are to be adapted to the marine environment. A clear understanding of the offshore conditions is essential for the design. Apart from the wind characteristics, the important factors are depth of water, wave conditions, see bed characteristics, proximity to electrical network, possibility of ice formation and effect on marine habitat. With the present day's technology, it is not feasible to build the farms at water depths over 40 m. Hence most of the farms are built near to the shore-probably within 10 km-where we have shallow water. Height of waves increases with the wind velocity and depth of seabed. Wave heights vary between 4 to 8 m at shallow bed where as at water depths between 16 to 29 m, the wave height could be as high as 16 to 20 m [14]. If site-specific measurements are not available, wave characteristics are to be modeled prior to the farm design. Similarly, possibilities of ice formation should also be taken into account.

Offshore turbines are to be guarded against corrosion with special surface finishing. Tower and nacelle are made airtight to avoid the exposure of sensitive components to salty air. Some designs are also provided with a dehumidification system. Due to the harsh marine environment, it may be difficult to reach the turbine frequently for maintenance. Hence, these high reliability- minimum maintenance turbines are designed to be remote controlled, as far as possible. Some of the modern turbines, tailor made for offshore use, are provided with special cranes for easy repair and maintenance of the blades and generator. Special platforms with cranes are also provided at the base for handling the components.

The distinct feature of offshore installations is its foundation, which are embedded in the seabed. The major types of foundations are gravity based foundation and piled foundation, the latter being the most common among offshore installations. The piled foundations can be mono-pile or tripod type as shown in Fig. 4.27. In tripod structures, load is distributed over a wider area. In gravity based structures, the tower and the turbine are fixed on a large and heavy mass of material. Due to the higher volume of material required, cost of the gravity based structures are higher than the piled foundations. Lattice and tubular towers can be used with both the piled and gravity based foundations, as shown in the figure.

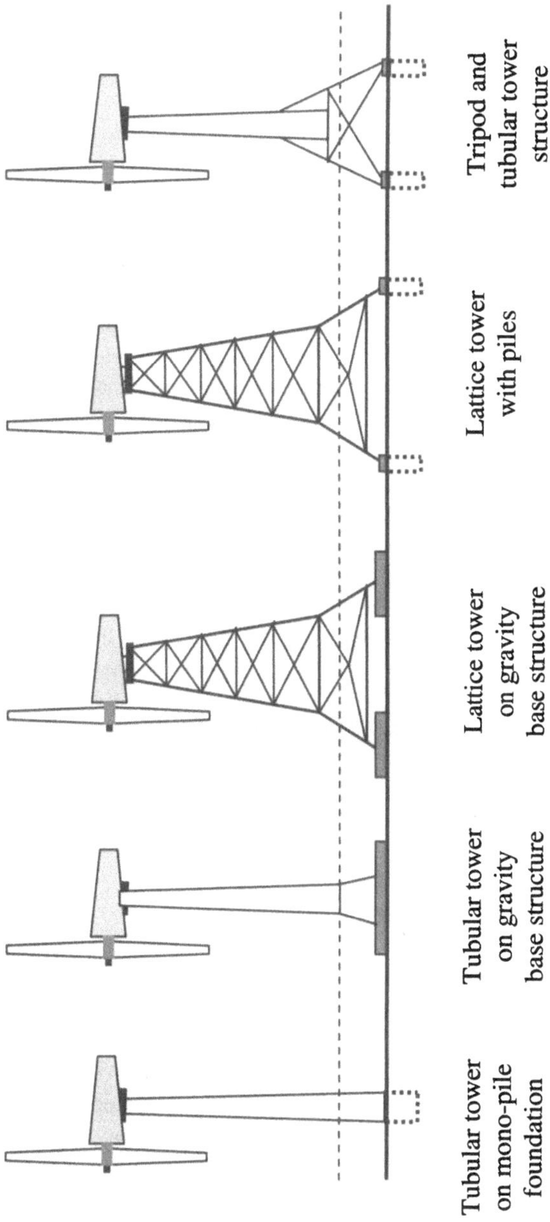

Fig. 4.27. Different foundations for offshore turbines

Another type of foundation, proposed for offshore turbines, is floating structure. Here, several turbines are placed together on a single platform. However, they are in the conceptualization phase and yet to be commercially adopted.

Owing to its special needs, offshore wind farms are costlier than onshore projects. As discussed in Chapter 7, the wind farm economics depends on several parameters like wind resource, characteristics of the location, access to grid, and prevailing energy market. At the current rate, an offshore wind farm approximately costs between \$ 14,000 and \$ 17,000 per kW installed capacity, which is around 30 per cent higher than the onshore installations. The major items adding to the installation costs are the foundation and grid connections. The foundation cost increases steeply with the water depth and wave height, while the distance from the shore determines the grid connection cost. In some cases, cost of support structure and grid connection went up to 40 per cent of the total project outlay. However, this high capital investment is often justified by the better energy yield from offshore farms.

Although some of the major environmental concerns of the land based wind farms are not relevant to offshore projects, they also have some environmental problems. Construction activities for installing offshore turbines may affect the marine habitats. This include the transportation of turbine and its components, construction of piles and other structures for the foundation, laying and burying of the cables and usage of chemicals and oils for the construction. It should be noted that, all these installation activities are usually completed within six months. The operational phase does not considerably affect the marine life. Though the operating noise from wind turbines may travel underwater, these low frequency waves are not audible to many marine species. The major environmental concern during the operation of offshore project is the bird collision. This is a site specific issue and, as in case of land based projects, areas of intensified avian activities should be avoided. European Commission has suggested a detailed procedure for the design of offshore wind farms [12].

4.4 Wind pumps

One of the classical applications of wind energy is water pumping. Several thousands of wind pumps are still in use in interior areas of America and Australia, catering to the water needs of crops and livestock. Tremendous scope for such machines do exist in many parts of the developing world, where grid connected power supply is not readily extendable due to technical and economic constraints. Gasoline, diesel or kerosene engines are being used to energize water pumps in these areas. Wind pumps are found economically competitive with these options, even at areas of moderate wind. Hence, with proper improvement in the technology and right policy initiatives, use of wind pumps is expected to continue in the coming years.

Wind pumps can broadly be classified as mechanical systems and electrical systems.

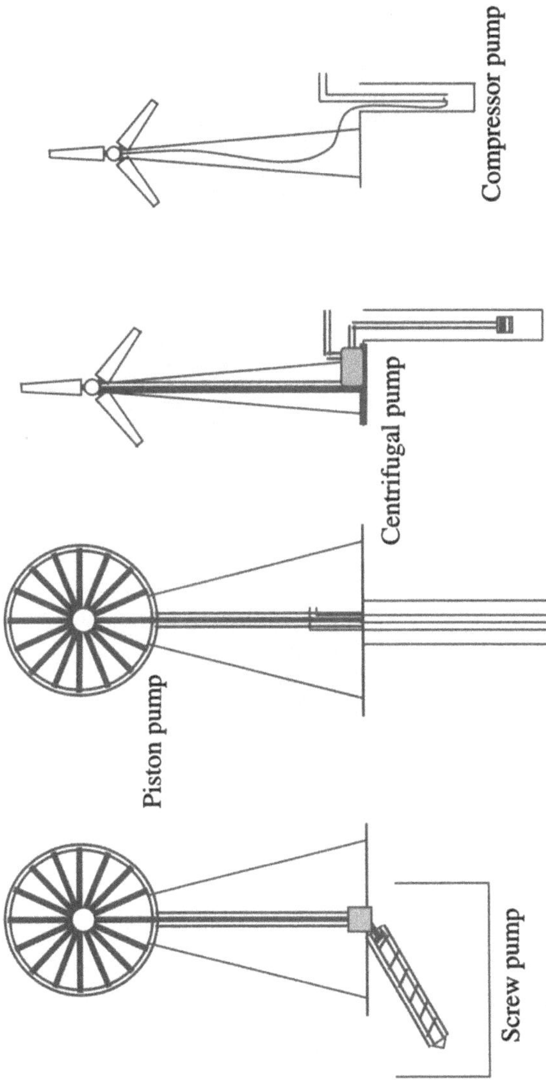

Fig. 4.2.8. Different pumps coupled mechanically to wind rotors

Fig. 4.29. A wind driven piston pump (Author: Pollinator, source: Wikipedia, http://wikipedia.org. GNU Free Documentation License applies to this image)

In mechanical wind pumps, the shaft power developed by the rotor is directly used to drive the pump. On the other hand, in electrical wind pumps, wind energy is first converted to electricity, which is then used to energize the pump. Mechanical wind pumps can further be categorized as systems with positive displacement and roto-dynamic pumps. Various types of pumps like the screw pump, piston pump, centrifugal pump, regenerative pump and compressor pump are being used in mechanical wind pumping option. Some of these configurations are shown in Fig. 4.28. Roto-dynamic pumps-mainly the centrifugal- are used with the electrical system.

4.4.1 Wind powered piston pumps

Positive displacement piston pumps are used in most of the commercial wind pumps. Constructional features of such a system are shown in Fig. 4.30. The system consists of a high solidity multi-vane wind rotor, drive shaft, crank, connecting rod and a reciprocating pump. Rotary motion of the windmill rotor is translated to reciprocating motion of the connecting rod by the crank. The connecting rod operates the pump's piston up and down through the cylinder during its strokes. Two check valves, both opening upwards, are fitted on the piston and the bottom of the pump. These valves allow the flow only in upward direction.

When the connecting rod drives the piston in the upward direction, the piston valve is closed and thus the water column above the piston is lifted up, until it is delivered out through the discharge line. At the same time, suction is created

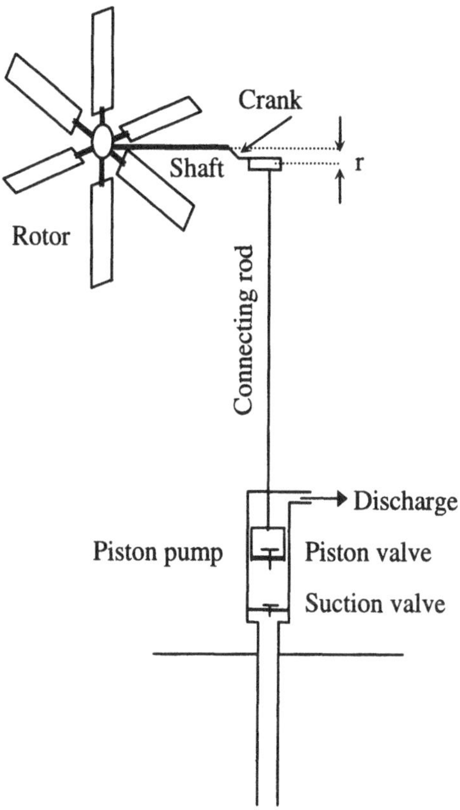

Fig. 4.30. Wind driven piston pump

below the piston, which causes the suction valve to open and thus fresh water from the well enter into the space below. During the downward stroke, the piston valve is opened and the suction valve is closed. The water collected below the piston thus enters into the space above, through the piston valve. These cycles are repeated resulting in pulsating sinusoidal water discharge from the system.

The volume of water discharged during one delivery stroke is given by the product of inner area of the cylinder and the height through which the water column is displaced during a stroke. Thus, if d is the inner diameter of the pump cylinder and s is the stroke length (distance between the extreme lower and upper positions of the piston) then, theoretically the volume of water pumped per discharge stroke is given by

$$V_s = \frac{\pi}{4} d^2 s \tag{4.8}$$

From the figure, we can see that

$$s = 2r \tag{4.9}$$

where r is the crank length. As the pump delivers one discharge per revolution of the wind rotor, the discharge is given by

$$Q = \eta_V \frac{\pi}{2} d^2 \; r \; N \tag{4.10}$$

where η_V is the volumetric efficiency of the pump and N is the rotational speed of the driving wind rotor. Usually, the volumetric efficiency of piston pumps is quite impressive, typically higher than 90 per cent. The power requirement of the pump (P_H) for a discharge Q may be estimated by

$$P_H = \frac{\rho_w \; g \; Q \; h}{\eta_p} \tag{4.11}$$

where ρ_w is the density of water, g is the gravitational constant, h is the total head against which the pump delivers water and η_p is the pump efficiency. Density of water, under standard ambient conditions, can be taken as 1000 kg/m^3. The pumping head includes the suction head, delivery head as well as the frictional head. Similarly, the pump efficiency takes care of various efficiencies involved in converting the mechanical shaft power to hydraulic power.

Example

Discharge of a wind powered piston pump at 6 m/s is 75 l/min. Pumping head is 20 m. Find out the rotor size of the wind pump. Power coefficient of the rotor at this velocity is 0.2 and the pump efficiency as 75 per cent.
The hydraulic power required by the pump for delivering 75 l/min (0.00125 m^3/s) against 20 m head is

$$P_H = \frac{1000 \times 9.81 \times 0.00125 \; \times 20}{0.75} = 327 \text{ W}$$

To develop 327 W at 6 m/s, the rotor area should be

$$A = \frac{327}{0.2 \times 0.5 \times 1.24 \times 6^3} = 12.2 \text{ m}^2$$

Thus the diameter of the rotor is 4 m.

Example

Determine the torque required for starting a wind driven pump with cylinder diameter 20 cm and stroke length 10 cm. Pumping head is 10 m.

As the stroke of the pump is 0.1 m, the crank arm is 0.05 m. Schematic view of the pumping system is shown in Fig. 4.31. When pumping starts, the piston is

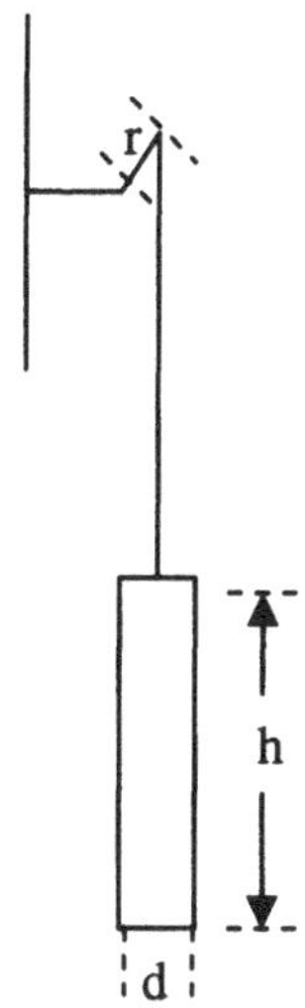

Fig. 4.31. Schematic view of the pumping system

at its extreme lower position. Hence, force acting on the rotor shaft due to the weight of water column is

$$W_W = \rho\, g\, h\, \frac{\pi \times d^2}{4} = \frac{\pi \times (0.2)^2 \times 10 \times 1000 \times 9.81}{4} = 3080\ \text{N}$$

This is acting at a distance of 0.05 m from the rotor shaft. Hence the torque required to start pumping is

$$3080 \times 0.05 = 154\ \text{N m}$$

4.4.2 Limitations of wind driven piston pumps

Mechanical coupling of a piston pump with wind rotor makes the system simple and cost effective. Several thousands of such pumps are installed at different parts of the world. However, the field performances of these units are not encouraging. The major reasons are:

- hysteresis behavior of the system due to its high starting torque demand
- mismatch between the characteristics of the rotor and the pump
- dynamic loading of the pump's lift rod.

These issues are discussed in brief in the following sections.

The hysteresis effect

In the above example we have seen that, the pump should overcome the weight of the water column acting on the piston to initiate the pumping action. To overcome this load, the wind rotor should develop a high torque at the starting point.

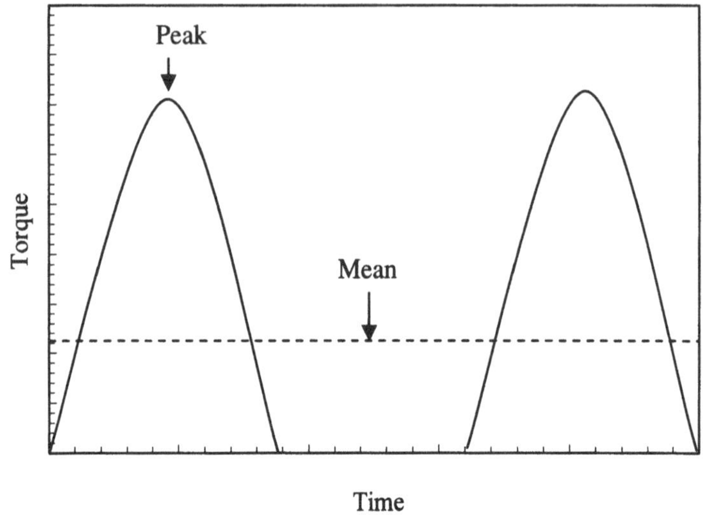

Fig. 4. 32. Torque characteristics of a wind driven piston pump

Thus, the system will start only at a wind, which is strong enough to develop this high initial torque needed by the pump. Taking the above example, a torque of 154 Nm is required to start the pump. Let us find out the wind velocity required to develop this torque. The torque developed by the rotor is given by

$$T_R = C_T \frac{1}{2} \rho A V^2 R \tag{4.12}$$

where C_T is the torque coefficient, A is the rotor area and R its radius. Wind driven piston pumps are usually coupled with high solidity multi-vane rotors showing high starting torque characteristics. Hence, let us take the torque coefficient of the rotor at start as 0.3. Considering the specifications of the pump in the above example, this torque can only be developed at a wind velocity of 5.7 m/s. This is obviously a high cut-in velocity for a system.

Now let us examine what happens once the system starts functioning. A single acting pump demand torque from the rotor only in its upward stroke. During the downward stroke, as the piston moves down freely through the water column without any water output, the torque demand is 0. Due to this reason, theoretical torque behavior of a piston pump is cyclic as shown in Fig. 4.32. However, practically, once the system starts pumping, the rotor functions as a flywheel due to its inertia and the system may experience only a more or less constant average torque as denoted by the dotted lines in the figure.

Hence, the torque behaviors of the system are different under starting and running conditions. The starting torque of the pump is approximately π times higher than the running torque. More precisely, considering the weight of the pump rod, the relationship between the starting and running torque is expressed as [4]

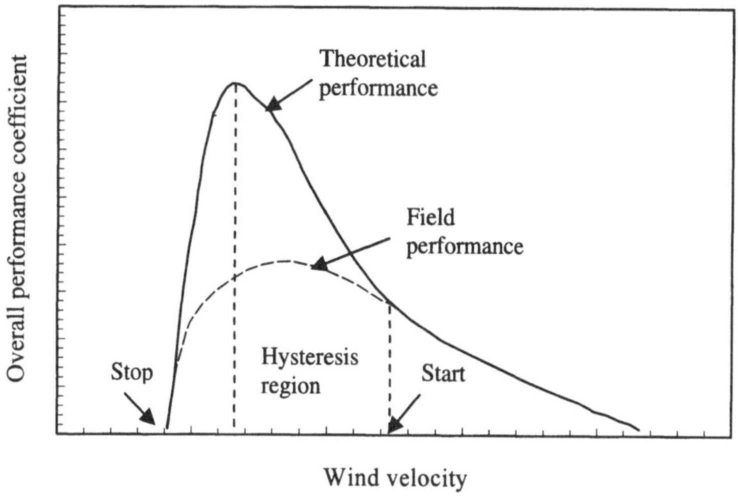

Fig. 4.33. The hysteresis effect

$$\frac{T_S}{T_M} = \pi \left(\frac{\eta_m}{\eta_V} \right) \left\{ 1 + \frac{F_m}{\rho_W \, g \, h \, A_p} \right\} \tag{4.13}$$

Here T_S is the peak torque demand of the pump, T_M is the mean demand, η_m and η_V are the mechanical and volumetric efficiencies, F_m is the force associated with the lift rod weight, and A_p is the cross-sectional area of the piston. Once the system starts pumping, it may continue to function even at velocities lower than cut-in velocity. Thus, the starting and stopping velocities of a wind driven piston pump are different. Such a system, when functioning under fluctuating wind regimes, may fall frequently into a hysteresis region between the starting and stopping wind speeds as shown in Fig. 4.33.

The wind speeds corresponding to the starting (V_{start}) and stopping (V_{stop}) of such systems are given by [27]

$$V_{start} = V_d \sqrt{\frac{\pi \, C_{qd}}{C_{qo}}} \tag{4.14}$$

and

$$V_{stop} = V_d \sqrt{\frac{p \, C_{qd}}{C_{q\max}}} \tag{4.15}$$

where V_d is the design wind velocity, C_{qd} is the design torque coefficient, C_{qo} is the torque coefficient at zero tip-speed ratio and $C_{q\,max}$ is the maximum torque developed by the rotor.

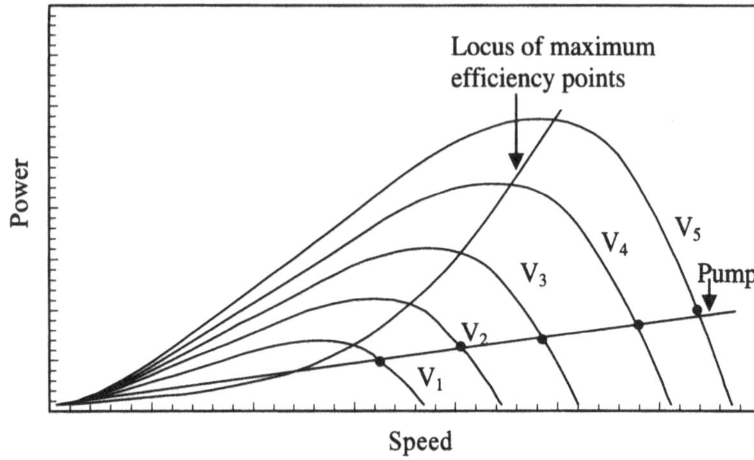

Fig. 4.34. Mismatch between the wind rotor and piston pump

At velocities between V_{Stop} and V_{Start}, the pump may either be running or standing still, depending on whether it was slowing down from running condition or stopped and required to be restarted again. This is the reason for the differences in the theoretical and field performances of the wind pump shown in the figure. Many commercial wind pumps do not practically start at low wind speeds as claimed by the manufacturers due to this high starting torque demand.

It should be noted that the design wind speed, which corresponds to the peek performance efficiency, lies in the hysteresis region. Functioning of the system around this best performance wind speed is not assured due to the hysteresis effect. This reduces the discharge of such wind pumps. Although a few methods like the use of loose fitting polyester vasconite pistons, piston leak holes, floating piston valves and double acting pumps are proposed for the smoother starting of such pumps, a satisfactory solution to this problem is yet to be derived.

Mismatch between the rotor and pump characteristics

Power demand of a piston pump increases linearly with its rotational speed. On the other hand, velocity-power relationship of a wind rotor is cubic in nature. Thus the rotor power rises steeply with rotational speed upto its peak and then declines with further increase in speed. The power-speed curves of a wind rotor and a piston pump are superimposed in Fig. 4.34. When such two systems are coupled, the combined system will settle down to work at the intersecting points of the power speed curves, which are denoted by thick dots in the figure.

It is evident that the working point of the combined system is away from the peak power points of the rotor. This is more apparent at higher wind velocities. Thus, though the piston pumps perform satisfactorily at lower wind velocities, it can use only a fraction of the power developed by the rotor at stronger winds. This wastage of power may be upto 90 per cent at velocities near to furling.

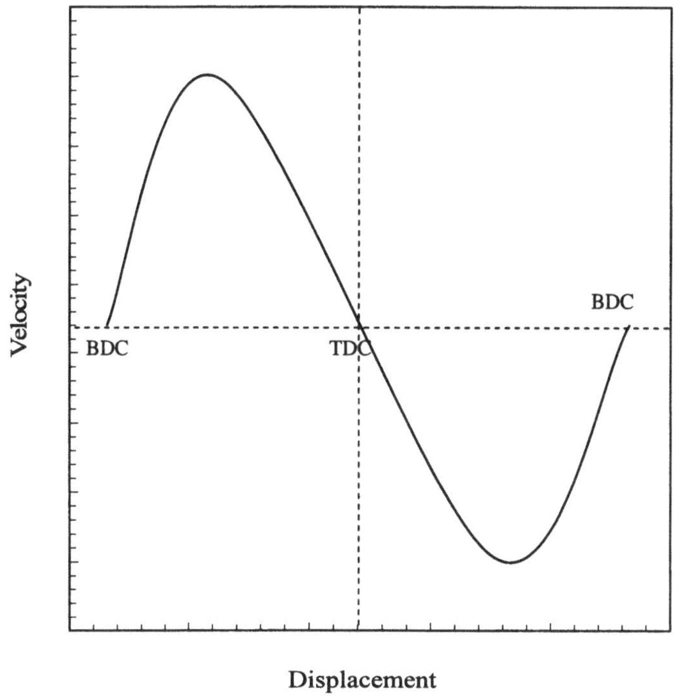

Fig. 4.35. Displacement-velocity diagram of an ideal piston pump

As a result of this poor matching between the wind rotor and piston pump, the overall system performance falls sharply, right after the peak performance has passed, as seen from the Fig. 4.33. This inability of the piston pumps to exploit the full potential of the rotor at higher wind speeds (which are more energy intensive as power is proportional to the cube of the wind speed) would be reflected as reduced system output.

Dynamic loading of the pump's lift rod

Recent designs of wind driven piston pumps have smaller rotors with lesser number of blades in comparison to the classical American mills. For simplicity and cost reduction, speed reducing back gears are also often eliminated from the system. For a given aerodynamic design, speed of the rotor is inversely proportional to its diameter and number of blades, thus preserving the tip-speed ratio. Consequently smaller rotors will tend to operate at higher speeds.

Displacement- velocity behavior of an ideal piston pump is illustrated in Fig. 4.35. The velocity of the piston varies in sinusoidal fashion as it is displaced from the bottom dead centre (BDC) to the top dead centre (TDC) and back to the BDC, after completing the suction and discharge strokes. During the upward movement of the piston, the velocity varies from zero (at BDC), reaches its peak at the middle of the stroke and returns back to zero (at TDC). This happens in the reverse direction in the next half cycle. When the operating speed is high, these changes in

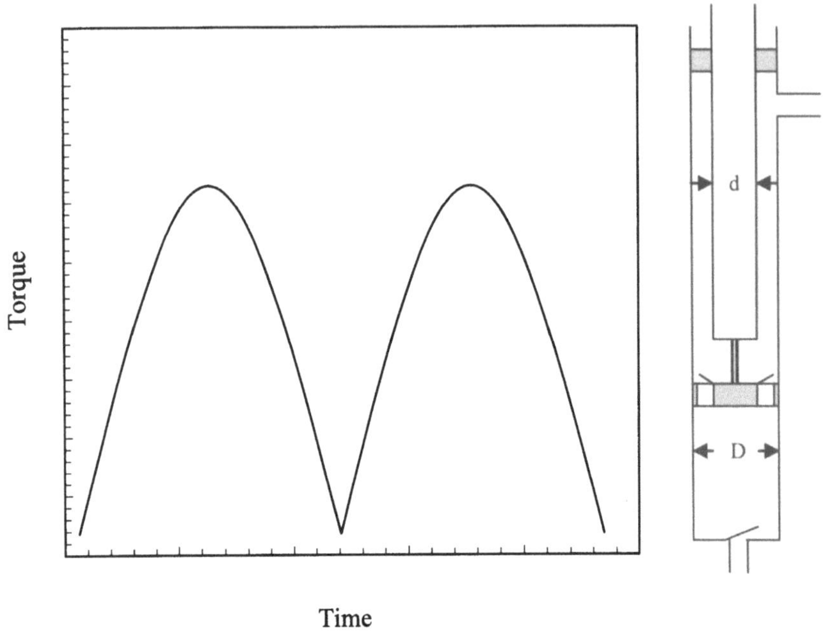

Fig. 4.36. Torque demand of a double acting pump

velocity from zero to the maximum and back to zero take place in a short span of time, which causes a very high acceleration/retardation. This high acceleration/retardation, when acts upon the mass of water column, causes undue forces on the lift rod. The peak lift rod loads goes upto 8 times that of the static load. This high dynamic load on the lift rods often causes it to fail. Breakage of lift rod accounts for more than 40 per cent of the total wind pump failures [3].

4.4.3 Double acting pump

Using double acting pumps is one possible way for reducing the high starting torque of wind driven reciprocating pumps. The double acting pump discharges during both its upward and downward strokes and thus offers a continuous flow. The torque demand is more uniform as shown in Fig. 4.36. For the same discharge per stroke, these pumps require only half of the starting torque in comparison with the single acting pumps. It can be shown that, for a double acting pump,

$$\frac{T_S}{T_M} = \frac{\pi}{2} \tag{4.16}$$

Due to this lower starting torque demand, these systems can be made to cut-in at lower wind speeds. Further, as the total flow is divided between the two strokes, acceleration effect is also minimized. This reduces the chances of lift rod failure. A clever design of the double acting pump, used in some commercial wind

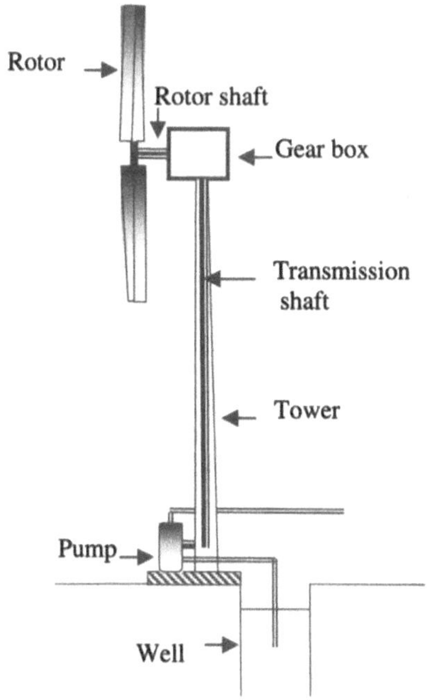

Fig. 4.37. Schematic view of a wind driven roto-dynamic pump [18]

mills is shown in Fig. 4.36. Here, cross-sectional area of the plunger is half that of the cylinder. Thus,

$$\frac{d}{D} = \frac{1}{\sqrt{2}} \tag{4.17}$$

During the upward stroke, water above the plunger is lifted and discharged. During the downward stroke water, equivalent to the volume of the plunger, is displaced and discharged. As the plunger cross section is half of the cylinder cross-section in area, the pump discharges equal amount of water in both upward and downward strokes, delivering an almost continuous flow. Wind driven double acting pumps cannot be used for deep well pumping (> 20 m) due to the danger of lift rod buckling during the downward stroke.

4.4.4 Wind driven roto-dynamic pumps

Wind pumping systems with roto-dynamic pumps were commercially manufactured by the Dutch even 65 years back. Several thousands of such machines were installed for low head applications like drainage. Schematic view of such a system is given in Fig. 4.37. As these pumps do not require high starting torque, low solidity rotors with 3 to 6 blades are usually used. Transmission components of this system consist of the rotor shaft, gear box and transmission shaft. Power from the

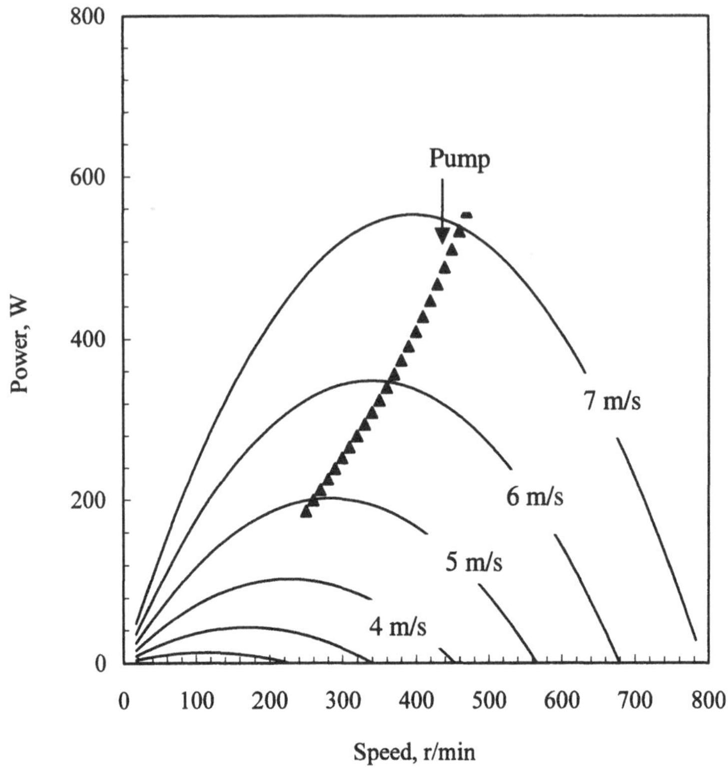

Fig. 4.38. Matching the wind rotor with roto-dynamic pumps

rotor shaft is transmitted to the transmission shaft using bevel gears. Another bevel gear at the bottom again converts the axis of power transmission as shown in the figure.

The major problems of wind pumps can be solved by using roto-dynamic pumps in the system. Starting and stopping wind speeds for a wind driven roto-dynamic pump are the same and thus, there is no hysteresis region during its operation. The power characteristic of these pumps matches well with that of wind rotors. For a roto-dynamic pump, power demand rises steeply with the rotational speed, after commencement of the flow. This makes it possible for the designer to match the pump at the peak efficiency points of the rotor. For example, the power speed behavior of a wind rotor at different wind velocities is superimposed with that of a roto-dynamic pump in Fig. 4.38. Power demand of the pump coincides well with the peak power point of the rotor. This shows that the pump can efficiently utilize the power developed by the rotor at all wind speeds. As the velocity of wind at a site varies extensively from time to time, better efficiency at broad zones of operation is advantageous.

In spite of its better performance, wind driven roto-dynamic pumps suffer from some inherent problems. First of all, they can only lift water from shallow depths. The practical suction head is limited to 7 m. For applications like irrigation and drainage, where large quantities of water are lifted from shallow depths, this suction head is quite sufficient.

Another concern is the gearing required for these systems. Most of the off-the-shelf roto-dynamic pumps like the centrifugal pumps are designed to run at very high speed in contrast with the low speed of the wind rotor. Hence complex gear trains are to be introduced in the system at the cost of transmission efficiency and economic viability. For example, a gear ratio of 1:24 is required for such systems with centrifugal pumps [1]. The gear requirement depends on the specific speed of the pump. The specific speed (N_{SP}), which basically defines the shape of the pump, is given by

$$N_{SP} = \frac{\eta_{pd}\sqrt{Q_d}}{(g\,H_d)^{0.75}} \tag{4.18}$$

where η_{pd} is the efficiency of the pump, Q_d is the pump discharge and H_d is the head, all at the design point. Once the specific speed of the pump used with the rotor is known, then the gear ratio (G) can be determined using the expression [1]

$$G = \sqrt{8\,\pi\left(\frac{\rho_W}{\rho_a}\right)\frac{1}{\eta_{pd}\,C_{Pd}}}\left(\frac{N_{sp}}{\lambda_d}\right)\left(\frac{gH_d}{V_d^2}\right)^{1.25} \tag{4.19}$$

where C_{Pd} is the design power coefficient, λ_d is the design tip speed ratio and V_d is the design wind velocity of the rotor. Another point of interest is the size of the pump in relation with that of the wind turbine. This is determined by the specific diameter of the pump (Δ), which is given by

$$\Delta = \frac{D_P\,(g\,H_d)^{\frac{1}{4}}}{\sqrt{Q_d}} \tag{4.20}$$

where D_P is the pump diameter. The ratio between the diameter of the turbine and pump can be given as

$$\frac{D_T}{D_P} = \sqrt{\frac{8}{\pi}\left(\frac{\rho_W}{\rho a}\right)\frac{1}{\eta_{pd}\,C_{Pd}}}\left(\frac{1}{\Delta}\right)\left[\frac{gH_d}{V_d^2}\right]^{0.75} \tag{4.21}$$

Effects of the specific speed and specific diameter on the gear ratio and diameter ratio of a typical wind driven roto-dynamic pump are shown in Fig. 4.39. The gear ratio increases with the specific speed whereas, the diameter ratio decreases with the specific diameter.

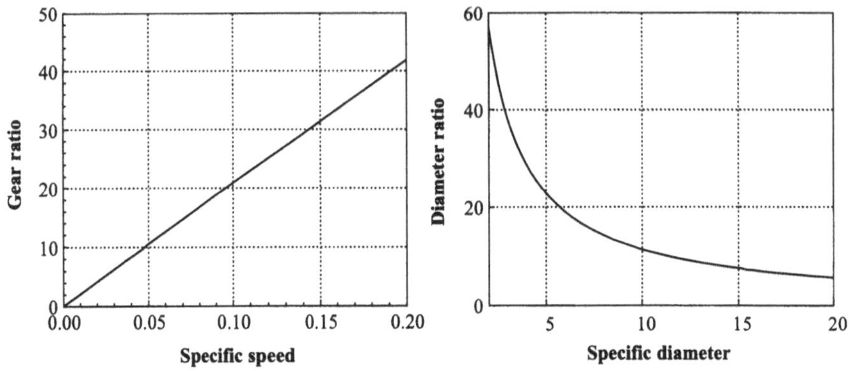

Fig. 4.39. Effects of the specific speed and specific diameter

Example

A centrifugal pump has specific speed of 0.095, specific diameter 5.5 and the peak efficiency 65 per cent. The pump is coupled with a 5 m wind rotor with maximum power coefficient of 0.4 at a tip speed ratio 2. If the pumping head is 5 m, find out the gearing required between the wind rotor and pump. Also calculate the size of the pump for optimum matching.
From the expression for the gear ratio, we have

$$G = \sqrt{8 \times \pi \times \left(\frac{1000}{1.24}\right) \times \frac{1}{0.65 \times 0.4}} \times \left(\frac{0.095}{2}\right) \times \left(\frac{9.81 \times 5}{6^2}\right)^{1.25} = 19.52$$

Similarly, the diameter ratio is

$$\frac{D_T}{D_P} = \sqrt{\frac{8}{\pi} \times \left(\frac{1000}{1.24}\right) \times \frac{1}{0.65 \times 0.4}} \times \left(\frac{1}{5.5}\right) \times \left[\frac{9.8 \times 5}{6^2}\right]^{0.75} = 20.37$$

For optimum coupling with 5 m turbine, diameter of the pump is 24.5 cm.

Limitations of coupling roto-dynamic pumps with wind turbines are evident from this example. Apart from high gear ratio, these systems require a bigger impeller to develop sufficiently higher peripheral velocity even at low shaft speeds. One possible solution for this problem is to use low specific speed roto-dynamic pumps. For example, if we couple a pump with a specific speed of 0.015 with the turbine described in the above problem, even with the lower pump efficiency of 0.5, the gear ratio can be reduced to 3.5. One such low specific speed pump, potentially be coupled with wind turbines is the regenerative pump. These pumps have flatter performance curves which make them specifically suitable for wind powered water pumping. A regenerative pump, exclusively designed for a wind pump is shown in Fig. 4.40 [19].

Fig. 4.40. The wind driven regenerative pump

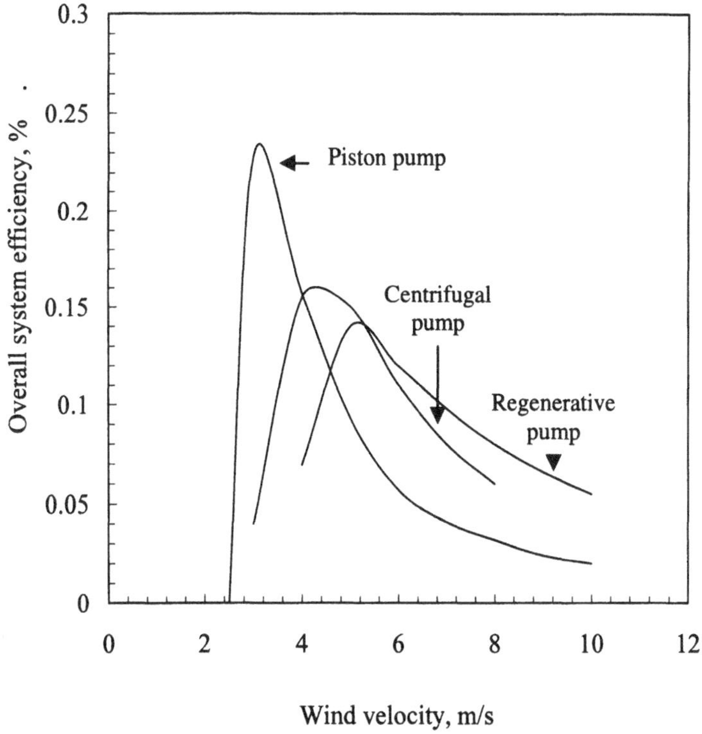

Fig. 4.41. Performance comparison of wind driven pumps

Performances of different pumps mechanically coupled with wind turbines are compared in Fig. 4.41. Though the piston pump showed a higher peak overall efficiency at lower velocities, its performance is poor at higher wind speeds. For example, at a wind speed of 10 m/s, efficiency of such systems drops to a low value of 2 per cent. Performance of roto-dynamic pumps are found better, especially at higher wind speeds. At 10 m/s, the regenerative pump could maintain an efficiency of 5.5 per cent. Thus, the roto-dynamic pumps offer better load to the rotor even at off-design wind velocities.

4.4.5 Wind electric pumps

In a wind electric pumping system (WEPS), a small and efficient wind electric generator energizes a roto-dynamic pump-mostly centrifugal-to deliver water in reliable and often cost effective way. Several industries around the world manufacture and distribute these systems which are extensively used for supplying water in remote areas.

Wind electric pumps have some distinct advantages over the mechanical wind pumps. As the electric power can be transmitted through cables, the wind turbine need not be installed just above the well as in case of a mechanical pump. This gives the flexibility of installing the turbine at a windy spot, for example at the top of a hill, where wind resource is strong. The pump can be placed at the valley where the water source is available. As WEPS do not require high starting torque, we can use fast running, low solidity rotors for these systems. This improves the efficiency and reduces the constructional cost of the rotor. Further, the output from the wind generator can be utilized for other applications like electrical lighting. The electrical transmission also results in lower maintenance and higher reliability.

A typical WEPS consist of a three bladed high speed rotor coupled with a permanent magnet generator. Most of the systems have custom built low speed generators driven directly by the rotor. If commercially available high speed generators are to be used, a gear box is required in the drive train to step up the rotor speed. Size of commercial systems ranges from 1 kW to 10 kW. Electricity from the generator is transmitted to the pump's motor through cables which may extend even upto 700 m. For deep well pumping, electro submersible centrifugal pumps can be used.

Electronic controllers are used to connect and disconnect the pump's motor from the wind turbine. These systems are highly reliable and reported to have 100 per cent availability for years, even without any attention. Although permanent magnet generators are common for WEPS, systems with induction generators are also available [20]. Here, the reactive power required for exiting the induction machines is delivered externally. The major drawback of WEPS is its high initial cost as it additionally requires the generator, motor and electronic control unit. In traditional wind pumps, the mechanical power available at the rotor shaft is directly used to lift the water. Whereas, in WEPS, we transform the shaft power into

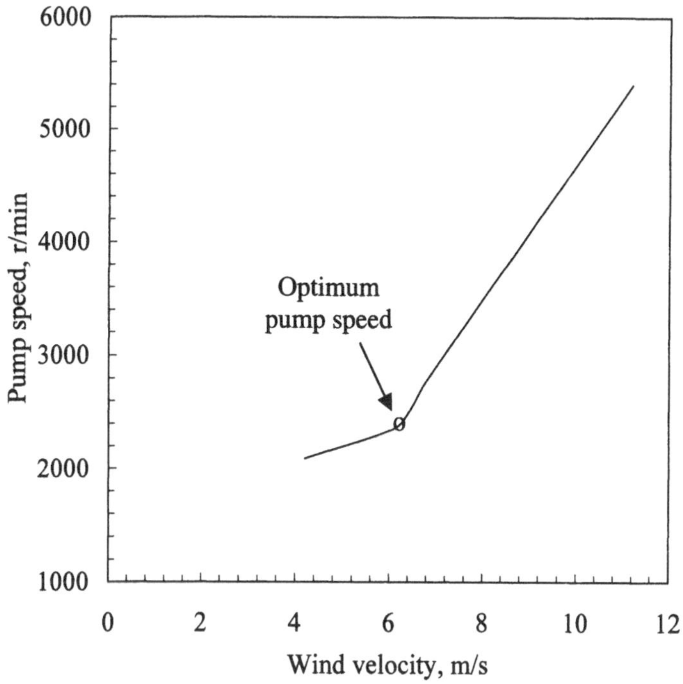

Fig. 4.42. Velocity-speed behavior of a WEPS

electrical form using the generator which is again converted back to mechanical power by the pump motor. This multistage transformation of power may cause drop in system efficiency. However, the better rotor-pump matching and flexibility and reliability of the system justify the use of WEPS for remote applications. WEPS are economical in areas where the annual average wind velocity is above 5 m/s.

Wind velocity-pump speed behavior of a typical WEPS is shown in Fig. 4.42. The system will start delivering water at a wind speed which is sufficient to overcome the hydraulic load on its shaft. The pump speed increases with the wind velocity. Point O corresponds to the pump speed at which the pump reaches its maximum efficiency. Further increase in the wind velocity results in a sharp increase in the pump speed. In the above example, at 11 m/s the pump speed is as high as 5200 r/min. Running the pump at this speed may cause the system to fail. Thus, the operating speed has to be restricted. This is one of the technical challenges faced by the wind electric pumping technology. Possible solutions are furling the rotor at higher wind speeds or more preferably restricting the turbine speed at constant level above a certain wind velocity. This in turn would demand a higher level of sophistication in the technology, which may not be economically justified for a small scale stand alone system.

References

1. Alam MM, Burton JD (1998) The coupling of wind turbines to centrifugal pumps. Wind Engineering 22 (5): 223-234
2. Brughuis FJ Advanced tower solutions for large wind turbines and extreme tower heights. http://www.mecal.nl
3. Burton JD, Davies DG (1994) Double acting wind pump systems for shallow wells. In: European Wind Energy Conference
4. Burton JD, Hijazin M, Rizvi S (1991) Wind and solar driven reciprocating lift pumps. Wind Engineering 15 (2): 95-108
5. BWEA Best practice guidelines for wind energy development. British Wind Energy Association, Kingsway, London
6. BWEA Prospects for offshore wind energy. (Report written for European Union, Altener contract XVII/4.1030/Z/98-395) British Wind Energy Association, Kingsway, London
7. Duque EPN, Johnson W, vanDam CP, Cortes R, Yee K (2000) Numerical predictions of wind turbine power and aerodynamic loads for the NREL phase II combined experiment rotor. In: AIAA/ASME Wind Energy Symposium, 38th AIAA Aerospace Sciences Meeting, AIAA-2000-0038, Reno, NV
8. Fitzwater LM, Winterstein SR (2001) Predicting design wind turbine loads from limited data: Comparing random process and random peak models. In: ASME Wind Energy Symposium, AIAA Aerospace Sciences Meeting, AIAA-2001-0046, Reno, Naveda
9. Garrad Hassan and Partners (1999) Electrodynamic braking of large wind turbines. ETSU W/45/00519/REP
10. Gipe P (2003) Wind turbine tower trends. http://www.Wind-Works.org
11. Hartnell G (2000) Wind on the system – Grid integration of wind power. Renewable Energy World, March – April
12. Kühn M, Harland LA, Bierbooms WAAM, Cockerill TT, Ferguson MC, Göransson B, van Bussel GJW, Vugts JH (1997) Integrated design methodology for offshore wind energy conversion systems. (Final report, JOR3-CT95-0087, European Commission, Non Nuclear Energy programme, Joule III)
13. Leclerc C, Masson C (1999) Predictions of aerodynamic performance and loads of HAWTS operating in unsteady conditions. In: ASME Wind Energy Symposium, 37th AIAA Aerospace Sciences Meeting and Exhibit, AIAA-99-0066, Reno, NV
14. Lindley D, Simpson PB, Hassan U, Milborrow DJ (1980) Assessment of offshore siting of wind turbine generators. In: Third International Symposium Wind Energy Systems, Copenhagen. BHRA, Cranfield
15. Malcolm DJ (2004) WindPACT rotor design study: hybrid tower design. NREL/SR-500-35546, National Renewable Energy Laboratory, Colorado

16. Manuel L, Veers PS, Winterstein SR (2001) Parametric models for estimating wind turbine fatigue loads for design. AIAA-2001-0047: 1-12
17. Mason KF (2004) Wind energy: Change in the wind. Composites technology, Ray Publishing
18. Mathew S, Pandey KP (2003) Modelling the integrated output of wind-driven roto-dynamic pumps. Renewable Energy 28: 1143-1155
19. Mathew S, Pandey KP, Burton JD (2002) The wind-driven regenerative water-pump. Wind Engineering 26 (5): 301-313
20. Miranda MS Lyra ROC, Silva SR (1999) An alternative isolated wind electric pumping system using induction machines. IEEE Transactions on Energy Conversion 14 (4): 1611-1616
21. Niederstucke B, Anders A, Dalhoff P, Grzybowski R. Load data analysis for wind turbine gearboxes. http://www.daneprairie.com
22. Parsons B, Milligan M, Zavadil B, Brooks D, Kirby B, Dragoon K, Caldwell J (2003) Grid impacts of wind power: A summary of recent studies in the United States. In: European Wind Energy Conference, Madrid, Spain
23. Rajagopalan RG, Fanucci JB (1985) Finite difference model for the vertical axis wind turbines. J Propulsion and Power 1: 432-436
24. Sorenson NN, Hansen MOL (1998) Rotor performance predictions using a Navier-Stokes method. In: ASME Wind Energy Symposium, 36th AIAA Aerospace Sciences Meeting and Exhibit, AIAA-98-0025, Reno, NV
25. Stiesdal H (1999) The wind turbine - components and operation. Bonus Energy A/S
26. Valpy B (2002) 2 MW wind turbine rotor development. ETSU W/45/00552/00/REP, URN 02/1433, NEG Micon Rotors Ltd.
27. van Meel J, Smulders P (1987) Some results from CWD's test fields: Are the IEA recommendations sufficient for windmills driving piston pumps? Wind Engineering 11 (2): 89-106
28. Westermann-Friedrich A, Zenner H, FVA-Merkblatt (1999) Zählverfahren zur Bildung von Kollektiven aus Zeitfunktionen, Frankfurt
29. Whale J, Fisichella CJ, Selig MS (1999) Correcting inflow measurements from HAWTS using a lifting surface code. In: ASME Wind Energy Symposium, 37th AIAA Aerospace Sciences Meeting and Exhibit, AIAA-99-0040, Reno, NV
30. Xu G, Sankar L (1999) Computational study of horizontal axis wind turbines. In: ASME Wind Energy Symposium, 37th AIAA Aerospace Sciences Meeting and Exhibit, AIAA-99-0042, Reno, NV

5 Performance of wind energy conversion systems

For the efficient planning and successful implementation of any wind power project, an understanding on the performance of the Wind Energy Conversion System (WECS), at the proposed site, is essential. The major factors affecting the power produced by a WECS are (a) the strength of the wind spectra prevailing at the site and its availability to the turbine (b) the aerodynamic efficiency of the rotor in converting the power available in the wind to mechanical shaft power and (c) the efficiencies in manipulating, transmitting and transforming this power into the desired form. Hence, assessment of the performance of a WECS is rather a complex process.

Wind is a stochastic phenomenon. Velocity and direction of wind at a location considerably vary from season to season and even time to time, as we have discussed in Chapter 3. Hence, apart from the strength of the wind spectra, its distribution also has significant influence on the performance of the WECS. Further, the characteristic operational speeds of a wind machine should match well with the prevailing wind spectra to ensure the maximum exploitation of the available energy. In this chapter, we will define the performance of different WECS by integrating these site parameters with the machine characteristics.

Several types of wind machines with the same rated capacity but different operating characteristics may be available in the market. The designer of a wind energy project should judiciously choose the turbine optimally matching with his site to maximise the energy production. Hence, the criteria for selecting the 'right machine for the right site' will also have to be included in our discussions.

As in Chapter 3, most of the expressions presented here cannot be solved through straightforward analytical methods. Readers who are not familiar with the numerical procedures may use the "WERA" programme provided with this book for solving the examples illustrated here. More help on WERA is available in Appendix 1.

5.1 Power curve of the wind turbine

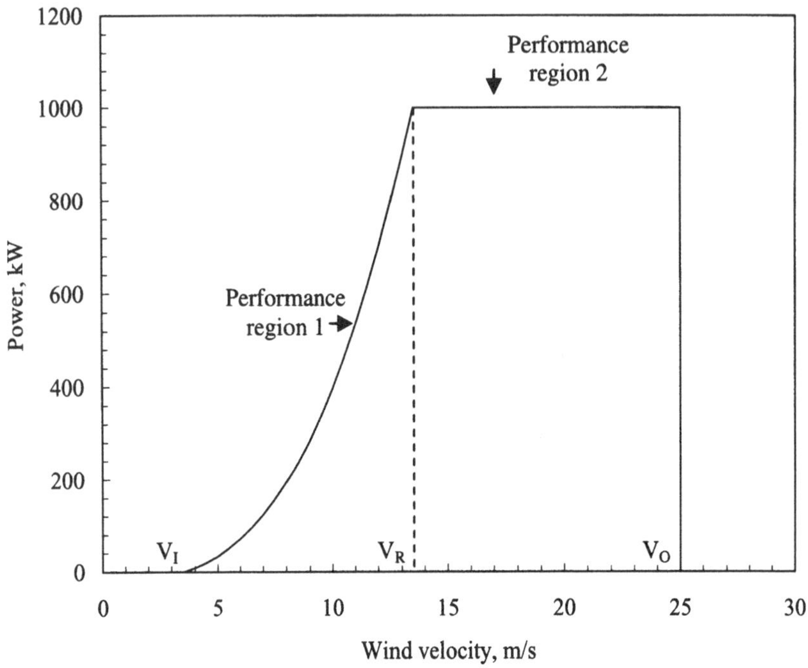

Fig. 5.1. Ideal power curve of a pitch controlled wind turbine

One of the major factors affecting the performance of a WECS is its power response to different wind velocities. This is usually given by the power curve of the turbine. The power curve of the machine reflects the aerodynamic, transmission and generation efficiencies of the system in an integrated form.

Fig. 5.1 shows the typical power curve of a pitch controlled wind turbine. The rated power of the turbine is 1 MW. The given curve is a theoretical one and in practice we may observe the velocity power variation in a rather scattered pattern. We can see that the important characteristic speeds of the turbine are its cut-in velocity (V_I), rated velocity (V_R) and the cut-out velocity (V_O). The cut-in velocity of a turbine is the minimum wind velocity at which the system begins to produce power. It should not be confused with the start-up speed at which the rotor starts its rotation. The cut-in velocity varies from turbine to turbine, depending on its design features. However, in general, most of the commercial wind turbines cut-in at velocities between 3 to 5 m/s.

Due to technical and economical reasons, the wind turbine is designed to produce constant power - termed as the rated power (P_R) - beyond its rated velocity . Thus, the rated velocity of a turbine is the lowest wind velocity corresponding to its rated power. Usually the system efficiency is maximum at V_R. From V_I, to V_R, the power generated by the turbine increases with the wind velocity. Between V_R and V_O, the turbine is restricted to produce constant power P_R

Table 5.1. Performance regions of a wind turbine

Velocity range	Power
0 to V_I	No power as the system is idle
V_I to V_R	Power increases with V
V_R to V_O	Constant power P_R
Greater than V_O	No power as the system is shut down

corresponding to V_R, irrespective of the changes in velocity. This power regulation is for better system control and safety. Hence P_R is the theoretical maximum power expected from the turbine. At wind velocities higher than V_O, the machine is completely shut down to protect the rotor and drive trains from damage due to excessive loading. Some times V_O is also termed as the furling velocity, as in the earlier sail type machines the canvas was rolled upto protect the mill from strong winds.

Hence, the turbine has four distinct performance regions as indicated in Table. 5.1. The power produced by the system is effectively derived from performance regions corresponding to V_I to V_R and V_R to V_O. Let us name these as region 1 and 2. The velocity-power relationship in the region 1 can be expressed in the general form

$$P_V = aV^n + b \tag{5.1}$$

where a and b are constants and n is the velocity-power proportionality. Now consider the performance of the system at V_I and V_R. At V_I, the power developed by the turbine is zero. Thus

$$aV_I^n + b = 0 \tag{5.2}$$

At V_R power generated is P_R. That is:

$$aV_R^n + b = P_R \tag{5.3}$$

Solving Eqs. (5.2) and (5.3) for a and b and substituting in Eq. (5.1) yields

$$P_V = P_R \left(\frac{V^n - V_I^n}{V_R^n - V_I^n} \right) \tag{5.4}$$

Eq. (5.4) gives us the power response of the turbine at wind velocities falling under the region 1. Obviously, the power corresponding to the region 2 is P_R.

Example

A 2 MW wind turbine has cut-in, rated and cut-out velocities 3.5 m/s, 13.5 m/s and 25 m/s respectively. Generate the power curve of the turbine.

Let us take the ideal velocity power proportionality, that is n = 3. Between 3.5 m/s and 13.5 m/s, the power is given by

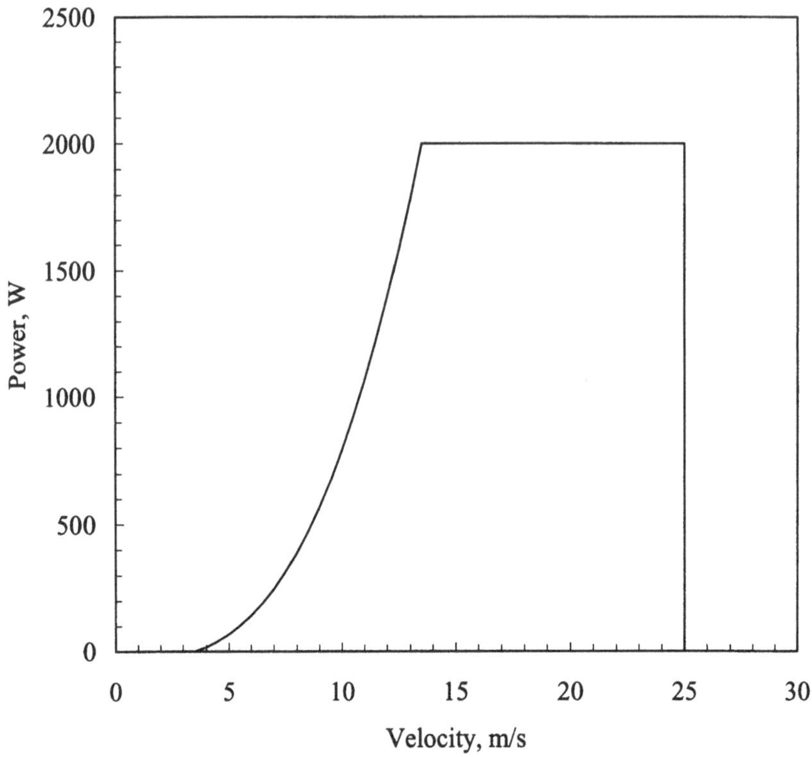

Fig. 5.2. Power curve of the wind turbine

$$P_V = 2000 \times \left(\frac{V^3 - (3.5)^3}{(13.5)^3 - (3.5)^3} \right)$$

For 13.5 <V<25, the turbine develops its rated power of 2,000 kW. When V>25, the turbine produces no power as it is shut down. The resulting power curves is shown in Fig. 5.2.

In practice, the power response of the turbine does not follow a perfect curve as shown above. Field observations are rather scattered (Fig. 5.3). This means that, for the same wind velocity V, the turbine appears to develop different powers as shown in the figure. This happens because we cannot precisely measure the strength of the wind as it passes through the turbine. The measurements are made from meteorological towers mounted slightly away from the turbine. Due to momentary fluctuations in the wind, velocity of the measured wind and the wind column actually interacting with the turbine may be different. Further, due to the variations in the air density, there may be differences in the power characteristics recorded during different days. Observations from wind farms indicate that even the same type of turbines installed at the same location may show different power characteristics [7]. It should also be noted that the power curve tells us the power expected from the turbine at an instantaneous wind velocity V and not for the velocity averaged over a period.

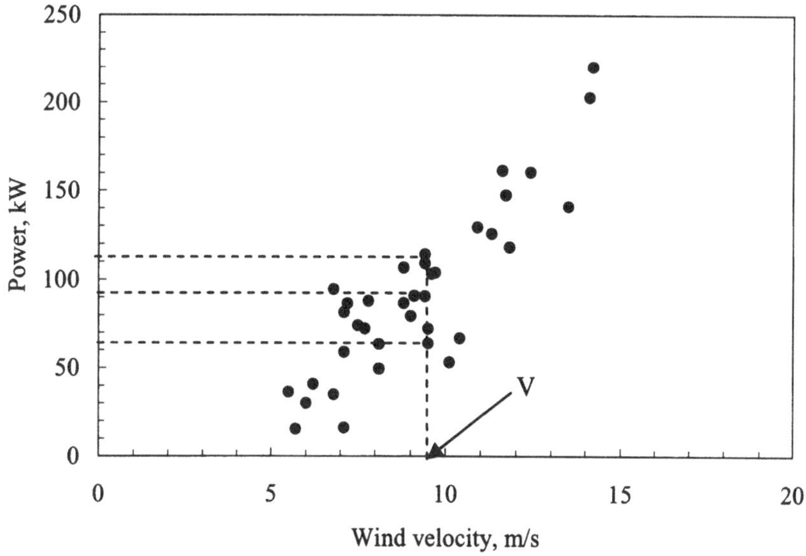

Fig. 5.3. Field performance of a 250 kW wind turbine

The velocity power proportionality has a profound effect on the power generated by the wind turbine. Ideally, variations in power with the velocity should be cubic in nature. However in practice, this can take any form such as linear, quadratic, cubic or even higher powers and its combinations [8,12]. For example, let us take the performance of the turbine considered in Fig. 5.3.

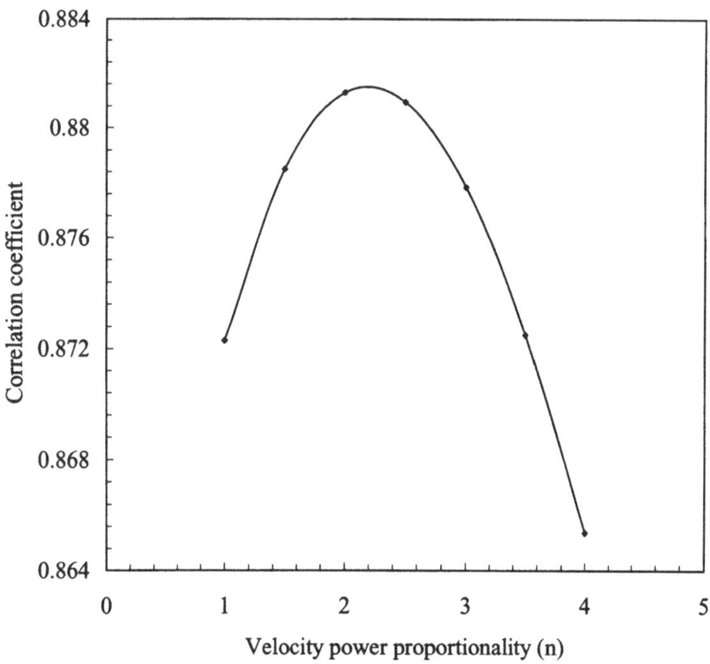

Fig. 5.4. Correlation between measured and generated performances

Velocity-power characteristics of this turbine for different n are simulated. The correlation coefficients between the observed and generated performances are plotted in Fig. 5.4. The correlation is highest for n = 2. Thus, this turbine shows a quadratic velocity-power relationship in its 1st performance region. Value of n varies from turbine to turbine. For Darrieus turbine, the velocity power relationship is generally linear [6], where as some two bladed horizontal axis designs show cubic velocity- power response.

5.2 Energy generated by the wind turbine

The factors influencing the energy produced by a WECS at a given location over a period are: (1) the power response of the turbine to different wind velocities (2) the strength of the prevailing wind regime and (3) the distribution of wind velocity within the regime. Energy corresponding to a certain wind velocity V is the product of the power developed by the turbine at V and the time for which this velocity V prevails at the site. The total energy generated by the turbine over a period can be computed by adding up the energy corresponding to all possible wind speeds in the regime, at which the system is operational. Hence, along with the power characteristics of the turbine, the probability density corresponding to different wind speeds also comes into our energy calculations.

This is expressed graphically in Fig. 5.5. In the figure, curve (a) represents the power curve of a commercial I MW turbine with 3.5 m/s cut-in, 15 m/s rated and 25 m/s cut-out velocities. Curve (b) represents the probability density function at a site with 8 m/s mean wind speed, assuming the Rayleigh distribution. The energy corresponding to different wind velocities are plotted in curve (c).

To find out the energy contributed by a wind speed of 10 m/s, the power and the probability density corresponding to 10 m/s are located from the curves (a) and (b). The energy corresponding to this velocity is indicated in the curve (c), which is basically the product of the power and the probability density for 10 m/s. Total energy available from this turbine at this site, over a period, can be estimated by integrating curve (c) within the limits of the cut-in and cut-out velocities and multiplying it with the time factor. Let us express these in mathematical terms in the following sections, adopting both the Weibull as well as Rayleigh distributions.

5.2.1 Weibull based approach

As we have seen in Chapter 3, if the wind velocity in a regime follows the Weibull distribution, its probability density function f(V) and cumulative distribution function F(V) are given by

$$F(V) = 1 - e^{-\left(V/c\right)^k} \tag{5.5}$$

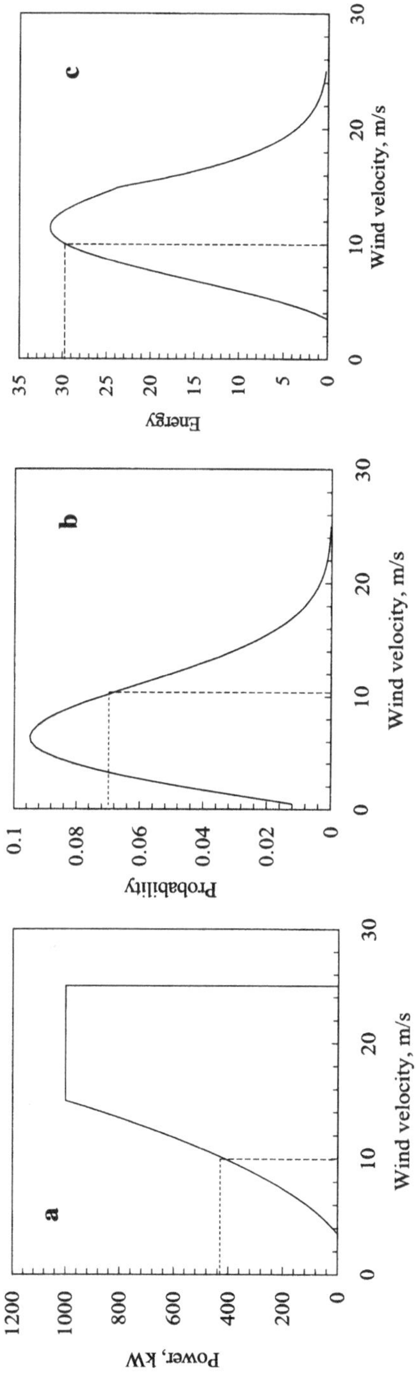

Fig. 5.5. Graphical representation of energy estimation

$$F(V) = 1 - e^{-\left(V/c\right)^k} \tag{5.5}$$

and

$$f(V) = \frac{dF}{dV} = \frac{k}{c}\left(\frac{V}{c}\right)^{k-1} e^{-\left(V/c\right)^k} \tag{5.6}$$

Here k is the Weibull shape factor and c is the scale factor. Methods to determine k and c are described in Chapter 3.

Referring to Fig. 5.1, let E_{IR} and E_{RO} be the energy developed by the system at its performance regions 1 and 2 respectively. Then E_{IR} can be computed by adding the energy corresponding to all possible wind velocities between V_I and V_R. Thus,

$$E_{IR} = T \int_{V_i}^{V_R} P_V \, f(V)\, dV \tag{5.7}$$

Similarly, E_{RO} can be expressed as

$$E_{RO} = T\, P_R \int_{V_R}^{V_O} f(V)\, dV \tag{5.8}$$

Substituting for P(V) and f(V) from Eqs. (5.4) and (5.6) in Eq. (5.7), we get

$$E_{IR} = P_R\, T \int_{V_I}^{V_R} \left[\frac{V^n - V_I^{\,n}}{V_R^{\,n} - V_I^{\,n}}\right] \frac{k}{c}\left(\frac{V}{c}\right)^{k-1} e^{-\left(V/c\right)^k}\, dV \tag{5.9}$$

$$= \left(\frac{P_R\, T}{V_R^{\,n} - V_I^{\,n}}\right) \int_{V_I}^{V_R} \left(V^n - V_I^{\,n}\right) \frac{k}{c}\left(\frac{V}{c}\right)^{k-1} e^{-\left(V/c\right)^k}\, dV \tag{5.10}$$

Now, let us introduce the variable X such that

$$X = \left(\frac{V}{c}\right)^k \tag{5.11}$$

Then,

$$dX = \frac{k}{c}\left(\frac{V}{c}\right)^{k-1} \quad \text{and} \quad V = c\, X^{\frac{1}{k}} \tag{5.12}$$

With Eq. (5.11), we have

$$X_I = \left(\frac{V_I}{c}\right)^k, \quad X_R = \left(\frac{V_R}{c}\right)^k \quad \text{and} \quad X_O = \left(\frac{V_O}{c}\right)^k \tag{5.13}$$

Thus, after simplification, E_{IR} can be expressed as

$$E_{IR} = \frac{P_R\, T\, c^n}{\left(V_R^{\,n} - V_I^{\,n}\right)} \int_{X_I}^{X_R} X^{n/k}\, e^{-X}\, dX - \frac{P_R\, T\, V_I^{\,n}}{\left(V_R^{\,n} - V_I^{\,n}\right)} \left[e^{-X_I} - e^{-X_R}\right] \tag{5.14}$$

Now consider the second performance region. From Eq. (5.8), E_{RO} may be represented as

$$E_{RO} = P_R\, T \int_{V_R}^{V_O} \frac{k}{c}\left(\frac{V}{c}\right)^{k-1} e^{-\left(V/c\right)^k}\, dV \tag{5.15}$$

As $\int f(V)\, dV = F(V)$, from Eqs. (5.4) and (5.5), Eq. (5.15) can be simplified as

$$E_{RO} = P_r\, T \left(e^{-X_r} - e^{-X_o}\right) \tag{5.16}$$

The total energy generated by the wind turbine, over a given period of time, is the sum of energy derived from the performance regions 1 and 2. Thus,

$$E_T = E_{IR} + E_{RO} \tag{5.17}$$

5.2.2 Rayleigh based approach

Rayleigh distribution is a simplified case of the Weibull distribution which is derived by assuming the shape factor as 2. Owing to its simplicity, this distribution is widely used for wind energy modeling. Under the Rayleigh based approach, the cumulative distribution and probability density functions of wind velocity are given by

$$F(V) = 1 - e^{-\frac{\pi}{4}\left(V/V_m\right)^2} \tag{5.18}$$

and

$$f(V) = \frac{\pi}{2}\frac{V}{V_m^2}\, e^{-\frac{\pi}{4}\left(V/V_m\right)^2} \tag{5.19}$$

As in the previous case, let us consider the performance regions 1 and 2 of the power curve (Fig. 5.1). Referring to Eq. (5.7), energy developed by the turbine (E_{IR}), while $V_I < V < V_R$, may be represented as

$$E_{IR} = \frac{P_R\, T}{V_R^n - V_I^n}\int_{V_I}^{V_R}\left(V^n - V_I^n\right)\frac{\pi}{2}\frac{V}{V_m^2}\, e^{-\pi/4\left(V/V_m\right)^2}\, dV \tag{5.20}$$

Substituting

$$X = \frac{\pi}{4}\left(\frac{V}{V_m}\right)^2,\ dX = \frac{\pi}{2}\frac{V}{V_m^2}\, dV \text{ and } V = 2\, V_m\left(\frac{X}{\pi}\right)^{1/2} \tag{5.21}$$

Eq. (5.21) implies that

$$X_I = \frac{\pi}{4}\left(\frac{V_I}{V_m}\right)^2,\quad X_R = \frac{\pi}{4}\left(\frac{V_R}{V_m}\right)^2 \quad and \quad X_O = \frac{\pi}{4}\left(\frac{V_O}{V_m}\right)^2 \tag{5.22}$$

Thus we can express E_{IR} as

$$E_{IR} = \frac{2^n\, V_m^n\, P_R\, T}{\pi^{n/2}\left(V_R^n - V_I^n\right)}\int_{X_I}^{X_R} X^{n/2}\, e^{-X}\, dX - \frac{P_R\, T\, V_I^n}{\left(V_R^n - V_I^n\right)}\int_{X_I}^{X_R} e^{-X}\, dX \tag{5.23}$$

This is further simplified as

$$E_{IR} = \frac{2^n\, V_m^n\, P_R\, T}{\pi^{n/2}\left(V_R^n - V_I^n\right)}\int_{X_I}^{X_R} X^{n/2}\, e^{-X}\, dX - \frac{P_R\, T\, V_I^n}{\left(V_R^n - V_I^n\right)}\left[e^{-X_R} - e^{-X_R}\right] \tag{5.24}$$

Now considering the performance of the system while $V_R < V < V_O$, we get

$$E_{RO} = T\ P_R \int_{V_R}^{V_O} \frac{\pi}{2} \frac{V}{V_m^2}\ e^{-\frac{\pi}{4}\left(\frac{V}{V_m}\right)^2}\ dV \tag{5.25}$$

With $\int f(V)\ dV = F(V)$ the above expression can further be simplified as

$$E_{RO} = T\ P_R \int_{X_R}^{X_O} e^{-X}\ dX \tag{5.26}$$

$$= T\ P_R \left[e^{-X_R} - e^{-X_O} \right] \tag{5.27}$$

Once we have expressions for E_{IR} and $E_{RO,}$ the total energy developed by the system E_T is obtained by adding E_{IR} and E_{RO}.

5.3 Capacity factor

Capacity factor is one of the important indices for assessing the field performance of a wind turbine. The capacity factor (C_F) of a WECS at a given site is defined as the ratio of the energy actually produced by the system to the energy that could have been produced by it, if the machine would have operated at its rated power throughout the time period. Thus

$$C_F = \frac{E_T}{T\ P_R} \tag{5.28}$$

The capacity factor reflects how effectively the turbine could harness the energy available in the wind spectra. Hence, C_F is a function of the turbine as well as the wind regime characteristics. Usually the capacity factor is expressed on an annual basis. Capacity factor for a reasonably efficient turbine at a potential site may range from 0.25 to 0.4. A capacity factor of 0.4 or higher indicates that the system is interacting with the regime very efficiently.

Information on the capacity factor of the turbine at a given site may not readily available during the initial phases of project identification. Under such situations, it is advisable to calculate the rough capacity factor (RC_F). This is basically deduced from the power curve of the machine, based on the average wind velocity at the site. From the power curve, we can locate the power corresponding to the average wind velocity. Dividing this power (P_{Vm}) by the rated power of the turbine, the rough capacity factor can be calculated. Thus,

$$RC_F = \frac{P_{V_m}}{P_R} \tag{5.29}$$

For example, let the average annual wind velocity at a site be 10 m/s. Consider that a wind turbine with the velocity-power response as in Fig. 5.2 is installed at this site. The power from the turbine at 10 m/s is 791 kW where as the rated power of the system is 2,000 kW. Thus the RC_F of the turbine at the site can be calculated as 0.39. We can calculate the approximate annual energy production by multiplying RC_F with the rated power and time period. Thus, in the above case, the annual energy production would be

$$2000 \times 0.39 \times 8760 = 6832800 \text{ kWh/year}$$

This is a very crude method. Wind energy potential of the site should be analysed in detail, following the methods discussed in the previous sections, before we proceed further with the project. It should be noted that, a system offering the highest capacity factor may not be always the most economic choice for a location. For example, consider a site having very strong wind spectra. Two wind turbines of the same rotor area, but different in generator size is being installed at this site. The turbine with smaller generator may show higher capacity factor at the site. However, the second system with the bigger generator may produce more energy at the site, in spite of its lower capacity factor. Thus the second turbine may prove to be more economical.

Example

A 2 MW wind turbine has cut-in, rated and cut-out velocities 3.5 m/s, 13.5 m/s and 25 m/s respectively. This turbine is installed at a site with wind data as given in Table 5.2. Estimate the performance of the wind turbine using the Weibull model.

Let us consider the ideal velocity-power proportionality, that is n = 3. The analysis is done using the "WIND TURBINE-WEIBULL" module of the WERA programme. The Weibull shape and scale factors for the site are 3.68 and 9.007 respectively. The total energy output from the turbine during the month is 378.9 MWh. The capacity factor is 0.26

Example

The Weibull factors k & c for four locations are given in Table 5.3. Estimate the performance of a wind turbine, with specification as in the above example, at these sites. Assume cubic velocity-power proportionality.

Table 5.2. Hourly wind data

	Days														
	1	2	3	4	5	6	7	8	9	10	11	12	13	14	15
1	9.6	9.6	6.9	4.6	4.1	8.3	7.4	3.5	3.8	4.6	5.5	5.5	4.9	5.5	4.9
2	9.7	8.0	7.2	4.1	3.6	6.6	5.8	4.7	4.1	4.7	5.0	4.4	5.0	6.4	5.8
3	8.9	8.7	8.4	4.5	4.5	7.3	6.2	4.5	6.5	5.1	4.3	4.8	5.1	8.4	7.6
4	9.3	9.1	6.8	4.1	3.8	7.8	4.1	2.4	6.0	4.3	4.6	4.6	5.5	8.0	8.2
5	9.9	9.4	6.3	4.6	4.4	9.1	3.3	4.4	6.6	5.2	6.1	4.4	5.5	8.1	7.8
6	9.9	8.8	7.8	5.5	4.6	8.0	3.2	3.2	7.4	4.3	3.8	4.1	5.2	7.4	6.7
7	11.2	10.6	10.0	7.2	5.6	8.9	5.6	4.7	8.2	5.0	5.9	6.2	7.8	8.6	7.5
8	11.3	10.5	10.8	9.1	6.6	9.3	8.0	6.3	8.8	5.5	5.8	6.8	7.1	9.6	7.4
9	11.7	10.9	10.6	9.6	8.1	10.6	9.8	6.7	9.8	6.7	6.7	8.1	9.2	9.2	8.1
10	13.8	12.1	10.7	11.0	10.2	11.6	9.9	8.5	9.3	8.5	7.1	9.1	9.6	11.0	10.2
11	13.6	10.8	9.7	10.5	9.4	11.6	11.6	8.6	8.0	8.6	7.2	8.3	9.7	9.4	9.7
12	14.3	11.5	10.4	11.2	10.1	12.6	9.8	9.0	8.7	8.7	6.2	9.3	9.8	9.8	8.7
13	13.9	11.1	10.3	9.4	10.7	12.4	7.5	9.7	10.1	8.4	4.9	8.4	9.4	8.4	9.4
14	13.3	9.8	11.2	9.7	11.8	12.4	6.8	9.4	10.2	7.4	6.4	8.4	9.4	9.0	9.4
15	13.6	10.5	10.5	9.4	12.4	11.2	4.7	8.1	9.7	7.9	7.6	8.6	9.7	7.8	10.4
16	13.1	9.1	10.4	8.6	12.4	10.6	3.4	10.4	10.1	7.7	8.4	9.4	9.6	8.4	9.6
17	11.1	8.4	9.4	9.4	10.7	10.1	3.1	10.0	7.3	7.1	8.4	8.4	9.4	8.4	9.2
18	11.1	8.9	9.4	9.1	8.4	9.4	1.8	8.4	6.9	6.7	6.7	7.9	8.4	7.8	8.4
19	10.6	9.4	9.7	7.7	6.3	10.1	1.9	7.4	5.8	5.7	6.3	7.2	7.9	7.3	6.7
20	10.8	8.7	8.7	7.4	6.6	10.4	2.5	8.5	6.4	6.4	5.7	6.4	6.1	7.1	6.6
21	10.7	8.5	8.4	6.3	7.0	10.4	3.1	5.9	5.4	4.6	5.6	5.4	7.1	6.3	6.7
22	11.2	8.4	6.4	5.7	6.8	9.4	2.6	5.4	4.5	5.1	5.0	6.7	5.7	5.1	6.3
23	8.5	7.9	5.9	4.4	6.7	9.4	4.7	5.1	4.7	5.7	5.7	6.4	6.5	5.4	6.7
24	11.6	7.9	5.4	4.2	8.4	9.7	5.1	6.5	4.7	5.4	5.0	6.7	6.4	4.2	6.6

	Days														
	16	17	18	19	20	21	22	23	24	25	26	27	28	29	30
1	7.1	4.6	4.6	5.8	6.3	7.1	10.2	6.6	6.0	1.6	5.8	4.4	5.2	10.8	5.8
2	6.6	5.0	4.7	5.3	5.0	6.1	9.4	5.8	5.0	1.1	4.1	3.6	5.0	10.8	5.0
3	6.5	5.4	6.2	5.4	3.2	8.0	10.1	6.8	6.2	2.3	5.1	3.2	3.7	11.5	5.9
4	5.8	4.9	5.8	4.3	3.2	8.2	10.2	6.8	6.0	3.0	5.2	3.0	4.1	11.0	5.5
5	6.6	5.5	6.3	7.7	4.4	9.1	10.2	7.4	6.0	3.6	4.6	2.4	4.4	11.0	6.1
6	7.4	5.5	6.8	8.0	4.9	7.4	10.7	5.8	6.1	4.1	2.7	3.2	2.7	10.2	6.3
7	7.8	6.6	6.4	8.1	6.2	6.7	11.1	7.0	8.1	4.2	3.1	5.0	5.3	11.7	7.0
8	8.5	6.8	6.3	6.6	6.0	7.4	11.3	6.8	7.7	6.0	5.7	3.0	4.6	12.1	6.8
9	9.0	8.4	5.6	8.1	6.5	7.0	11.5	6.7	6.7	6.5	6.5	4.2	4.0	12.0	6.7
10	9.6	9.9	8.8	9.3	6.3	8.8	11.0	8.8	6.3	5.7	7.1	6.8	6.0	11.6	6.8
11	9.4	10.0	8.9	7.8	7.8	8.0	11.9	6.1	5.5	4.4	8.9	5.8	8.0	11.6	7.8
12	9.0	10.4	9.8	6.5	8.2	7.3	11.2	5.1	6.0	5.7	9.3	5.7	11.2	11.7	8.7
13	9.3	10.2	9.3	6.8	8.1	5.9	10.5	5.8	3.9	8.4	6.1	11.1	10.9	6.4	6.9
14	9.4	10.4	8.0	5.7	8.9	5.8	8.3	6.1	4.3	6.4	4.5	13.1	10.1	5.7	5.4
15	9.7	10.7	5.4	4.8	8.8	5.8	9.5	6.5	4.8	6.8	4.2	12.4	11.0	3.4	3.7
16	9.4	9.6	6.1	3.5	5.7	9.6	9.7	5.3	3.2	4.6	4.4	11.9	11.3	6.9	6.9
17	9.3	6.5	6.2	3.1	6.4	9.4	9.3	4.4	2.1	6.4	3.7	12.1	10.1	6.4	6.9
18	9.2	5.7	8.9	6.5	7.9	8.7	9.5	6.4	2.6	8.4	4.5	11.7	9.4	6.8	6.5
19	8.4	5.4	7.6	6.1	8.4	11.3	9.1	7.5	3.4	6.9	5.8	11.4	9.4	5.1	5.6
20	8.1	6.0	5.7	6.7	9.2	11.4	9.8	8.0	5.1	5.4	5.7	11.9	10.1	6.4	5.8
21	6.4	6.1	5.9	5.9	8.4	10.7	9.0	8.4	5.3	6.2	5.2	10.4	8.7	6.2	6.3
22	4.7	4.0	6.7	5.7	8.2	11.5	9.1	6.4	6.7	5.7	5.9	10.8	6.9	5.4	5.7
23	4.8	2.7	6.8	6.0	7.9	11.7	8.4	5.4	6.8	5.4	4.6	11.0	6.4	5.9	5.7
24	4.3	3.4	6.4	5.6	7.2	11.3	7.6	5.0	7.4	4.3	5.0	11.0	5.8	4.3	4.7

Table 5.3. Weibull k and c for the locations

Location	1	2	3	4
k	2.61	3.35	2.93	2.31
c	8.73	7.92	11.50	6.98

The "WIND TURBINE-WEIBULL" module of WERA is again used for solving this problem. Results are given in Table 5.4.

Table 5.4. Performances of the wind turbine at the locations

Location	k	c	Energy, MWh/year	C_F
1	2.61	8.73	4574.84	0.26
2	3.35	7.92	3149.05	0.18
3	2.93	11.50	8500.30	0.49
4	2.31	6.98	2541.47	0.15

Example

Using the Rayleigh distribution, estimate the performance of the wind turbine described in the above examples at a site with wind speeds as given in Table 5.2. Assume cubic velocity-power proportionality.

The problem is solved using the "WIND TURBINE-RAYLEIGH" module of the WERA programme. The average wind velocity at the site is 7.38 m/s. With a capacity factor of 0.25, the energy generated by the turbine at this site over the month is 363.96 MWh. Comparing this with the results obtained under the Weibull based analysis (for the same turbine at the same wind regime), it is seen that the Rayleigh based calculations resulted in slightly lower performance values. This is because in the Rayleigh based analysis, we assumed k as 2, whereas the actual k for the site is 3. 68.

Example

The monthly mean wind velocities at a site are given in Table 5.5. Estimate the performance of the above wind turbine at this site, assuming quadratic velocity-power proportionality.

The "WIND TURBINE-RAYLEIGH" module is used for this analysis. Results are shown in Table 5.6.

Table 5.5. Monthly mean wind velocities

Month	V_m, m/s	Month	V_m, m/s	Month	V_m, m/s	Month	V_m, m/s
January	10.5	April	13.4	July	14.4	October	10.0
February	12.0	May	12.5	August	13.9	November	10.6
March	12.9	June	13.1	September	12.1	December	11.6

Table 5.6. Performance of the wind turbine at different months

No	Month	V_m, m/s	Energy, MWh	C_F
1	January	10.5	672.73	0.47
2	February	12.0	763.59	0.53
3	March	12.9	797.70	0.55
4	April	13.4	810.62	0.56
5	May	12.5	784.33	0.54
6	June	13.1	803.36	0.56
7	July	14.4	825.11	0.57
8	August	13.9	819.64	0.57
9	September	12.1	768.11	0.53
10	October	10.0	632.82	0.44
11	November	10.6	680.15	0.47
12	December	11.6	743.62	0.52

It should be noted that, we considered the power characteristics of a pitch controlled wind turbine for our analysis. These machines are very common among large-scale wind power projects. However, for the stall and yaw controlled turbines, the performance curves are different as we have seen in Chapter 4. The performance analysis presented above can also be applied for stall/yaw controlled systems, obviously with lower accuracy.

5.4 Matching the turbine with wind regime

By now, it is evident that performance of a WECS at a site depends heavily on the efficiency with which the turbine interacts with the wind regime. Hence, it is essential that the characteristics of the turbine and the wind regime at which it works should be properly matched. The capacity factor of the system can be a useful indication for the effective matching of wind turbine and regime. For turbines with the same rotor size, rated power and conversion efficiency, the capacity factor is influenced by the availability of the turbine to the prevailing wind.

In other words, the functional velocities of the turbine (V_I, V_R and V_O) should be chosen in such a way that, the energy available with the wind regime is exploited to its maximum level. This in turn would require that the turbines should be individually designed for each site so that these functional parameters can be defined according to the site characteristics. This is not practical.

Several wind turbines of different ratings and functional velocities are available in the market. A wind energy project planner can choose a system, best suited for his site, from these available options. Hence, it is important for him to identify the effect of these functional velocities- that is V_I, V_R and V_O - on the turbine performance at the given location and ensure that the turbine and the wind regime are working in harmony. The performance estimation methods discussed in the above sections can be used for such an analysis.

For example, consider a 2 MW turbine with 12 m/s rated velocity installed at the site described in Table. 5.2. Let us examine the effect of the cut-in

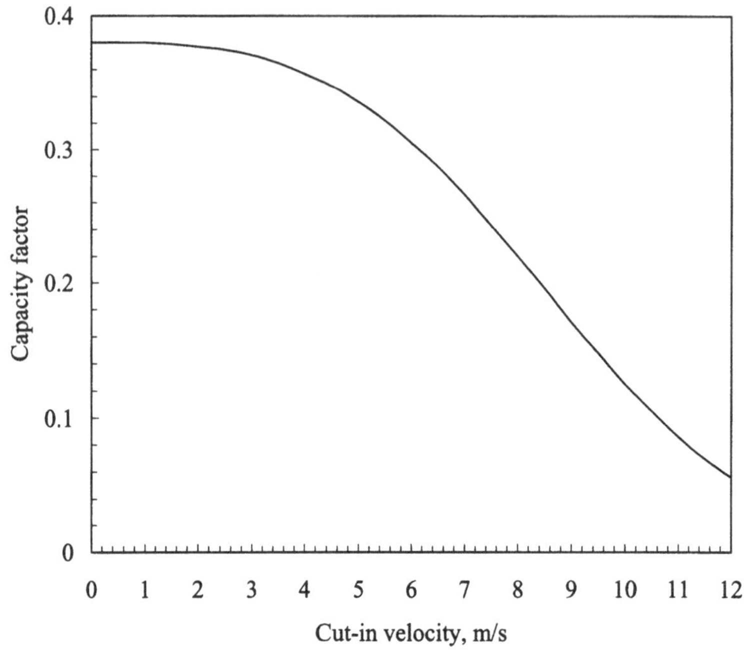

Fig. 5.6. Effect of cut-in velocity on the system performance

and cut-out velocities on the performance of the turbine at this site. The "WIND TURBINE-WEIBULL" module of the WERA programme can be used here. In-order to identify the effect of cut-in velocity, compute the capacity factors of the turbine by varying V_I at different levels. V_O is kept at a reasonably higher value for this analysis. The capacity factors thus obtained are plotted against the respective cut-in speeds in Fig. 5.6. Upto a velocity of 3.0 m/s, the cut-in velocity does not have any significant influence on the capacity factor. However, for cut-in velocities higher than 3.0 m/s, there is a noticeable decrease in the capacity factor. Similar procedure may be followed for identifying optimum V_O of the turbine. In this case, V_I is fixed and V_O is varied. Results are shown in Fig. 5.7. Effect of V_O on the system performance is prominent up to 16 m/s. With further increase in the cut-out velocity, the capacity factor is not improved considerably.

In the above procedure, we have identified the effect of V_I and V_O based on the site characteristics. However, the machine characteristics also have to be considered while choosing V_I and V_O for a system. The cut-in speed should be strong enough to overcome all the system losses. Considering the efficiencies of power transmission and generation, a higher cut-in speed, say 0.3 to 0.5 times the rated speed, may be suggested in some situations [13]. Similarly, capability of the system in sustaining extreme aerodynamic loads should be considered while fixing up V_O. Owing to engineering and economic reasons, the cut-out speed normally does not exceed $2V_R$ in most of the commercial designs.

This indicates that V_I and V_O are influenced by V_R. For a given rotor area and efficiency, the rated speed is directly correlated with the system's rated power. Now let us examine the effect of V_R on the capacity factor taking V_I = 3 m/s V_O = 16 m/s (Fig. 5.8). The capacity factor decreases with increase in V_R.

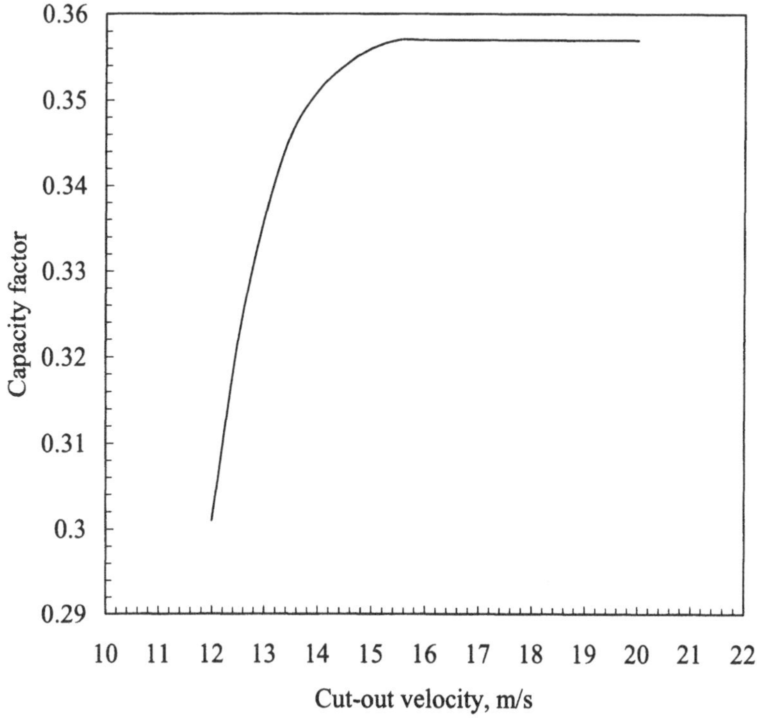

Fig. 5.7. Effect of cut-out velocity on the system performance

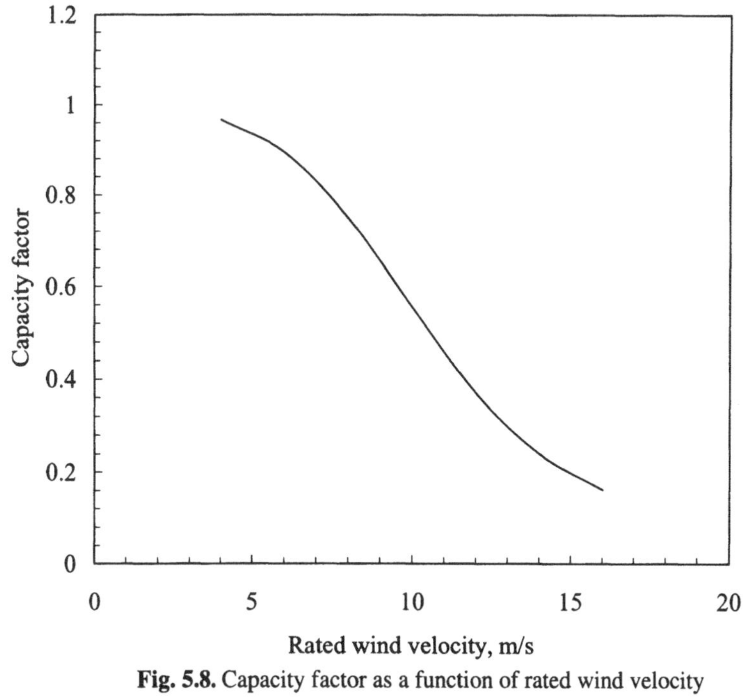

Fig. 5.8. Capacity factor as a function of rated wind velocity

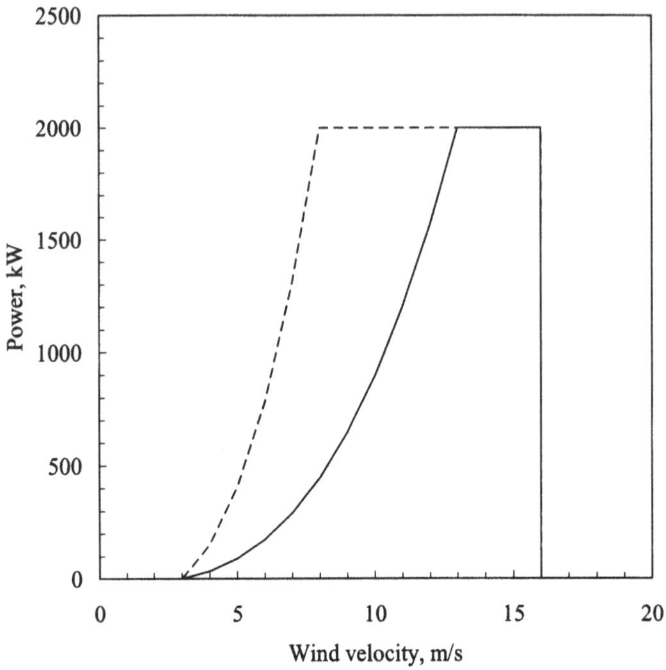

Fig. 5.9. Comparison of power curves for different rated wind speeds

The reason is evident from Fig. 5.9, in which the power curves of two similar systems, which differ only in the rated velocity, are compared. With the increase in V_R, area under the power curve reduces, which is finally reflected as the reduction in the capacity factor. However, for the same rated power, lower rated velocity will in turn demand a bigger rotor for the turbine. As the capital investment required for the WECS is proportional to the rotor area, this will increase the unit cost of energy produced.

For a given rotor size, increase in the rated velocity means increase in the rated power and thus the generator size (assuming that the efficiency is unchanged). If this higher rated velocity is justified by the strength and nature of the prevailing wind regime, this would in turn improve the energy production and thus reduce the cost of unit energy generated. At the same time, if the rated wind speed is too high for the regime, the system will seldom function at its rated capacity.

The above discussions are hypothetical and meant only to demonstrate the effect of V_I, V_R and V_O on the turbine performance. Unless under special situations, it is not practical to design wind turbines for a specific site. So let us look into this problem in a more practical point of view.

Several wind machines of the same power class but differing in performance curves may be commercially available. Designer of the wind energy project often chooses a system from these available options for his site. Selection of the 'right machine for the right site' plays a major role in the success of the project. Depending on the Weibull scale and shape factors, it is possible to identify V_R suitable for a particular wind regime [4,5,13]. It is suggested that the rated wind speed may be about twice the average wind speed for regimes with k = 2. In trade winds with higher k, V_R may be 1.3 times V_m.

The turbine performance models discussed in sections 5.2 and 5.3 can be used to choose the turbine which is most suited for a given wind regime. This is shown in the example below.

Example

Specifications of four commercial wind turbines are given in Table 5.7 Choose the turbine best suited for the wind regime described in Table. 5.2.

Table 5.7. Specifications of the wind turbines.

Specification	Wind Turbine			
	1	2	3	4
Rated Power (kW)	250	250	250	250
Rated velocity (m/s)	13.5	13	15	15
Cut-in velocity (m/s)	3.5	4	3	4
Cut-out velocity (m/s)	25	25	25	25

Let us consider the Weibull based analysis with ideal velocity power proportionality. Performances of the turbine at the site are estimated using the "WIND TURBINE-WEIBULL" module of the WERA programme. The resulting capacity factors of the systems are compared in Fig. 5.10

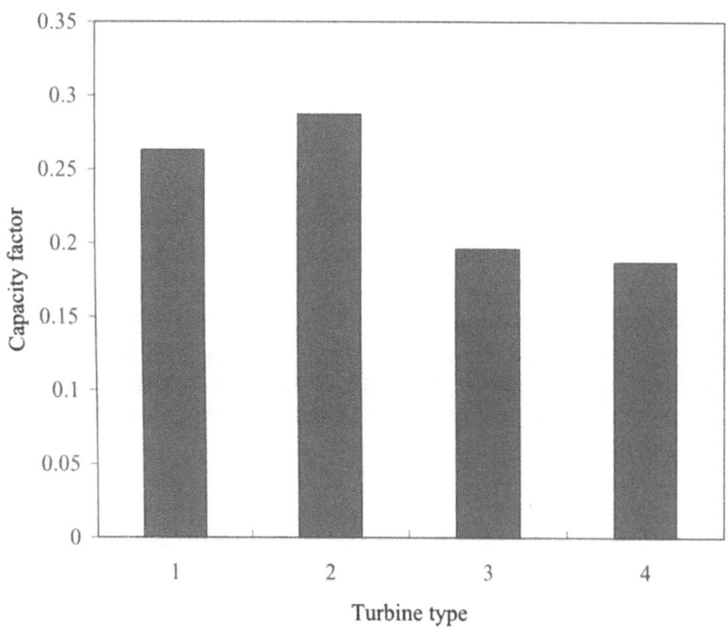

Fig. 5.10. Comparative performances of the wind turbines

Although the rated powers of all the turbines are equal, turbine-2 with its highest capacity factor offers the best performance in this wind regime. Let us compare the turbines 2 and 4 in particular as they are similar in all other respects, except for the rated wind speed. The capacity factor for turbine-4 is only 0.187 against 0.287 for the second system. It should be noted that the average wind velocity at the site is 7.4 m/s, which makes the V_R/V_m ratio 1.76 for the 2nd turbine and 2.03 for the 4th one. As we have seen, V_R/V_m is optimum around 1.3 for such locations.

5.5 Performance of wind powered pumping systems

One of the potential applications of wind energy is for water pumping, especially in remote rural areas, where grid connected power supply is not readily available. Several thousands of such systems are still functional in different parts of the world. These pumps can have either mechanical or electrical power transmission. Constructional features of these systems are discussed in Chapter 4.

Performance of a wind pump in a given wind regime is governed by (1) the behavior of its two major components viz. the wind rotor and the pump. (2) Effective interaction of the rotor-pump combination with the wind regime. Several methods are proposed to estimate the wind pump performance under fluctuating conditions of wind regimes. In the most common approach, a constant overall efficiency is assumed throughout the working range of the system. Here, the pump and wind regime characteristics are not considered. Though this sounds simple and handy, this approach may result in overestimating the capacity.

Some elaborate methods, taking the rotor and the pump characteristics into account, were also suggested [1,14]. However, the interaction of the rotor-pump combination with the wind regime is overlooked in these models. Hence, these models give only the instantaneous performance of the system in terms of expected discharge at a given wind velocity.

We will follow an integrated approach incorporating the characteristics of the rotor, pump and the wind regime, for defining the system performance. All the common types of systems viz. (1) wind driven piston pumps with mechanical transmission, (2) wind driven roto-dynamic pumps with mechanical transmission and (3) wind driven roto-dynamic pumps with electrical transmission, are considered. Most of the wind pumps are to be installed in remote rural areas. Wind data available for such locations are in the form of seasonal averages (for example daily or even monthly mean wind velocity). Hence Rayleigh distribution is used for our analysis in the following sections.

5.5.1 Wind driven piston pumps

A simple multi-bladed windmill mechanically coupled to a piston pump without any back gearing and power regulation is considered here. Such pumps are popular in many parts of the world, especially in developing countries.

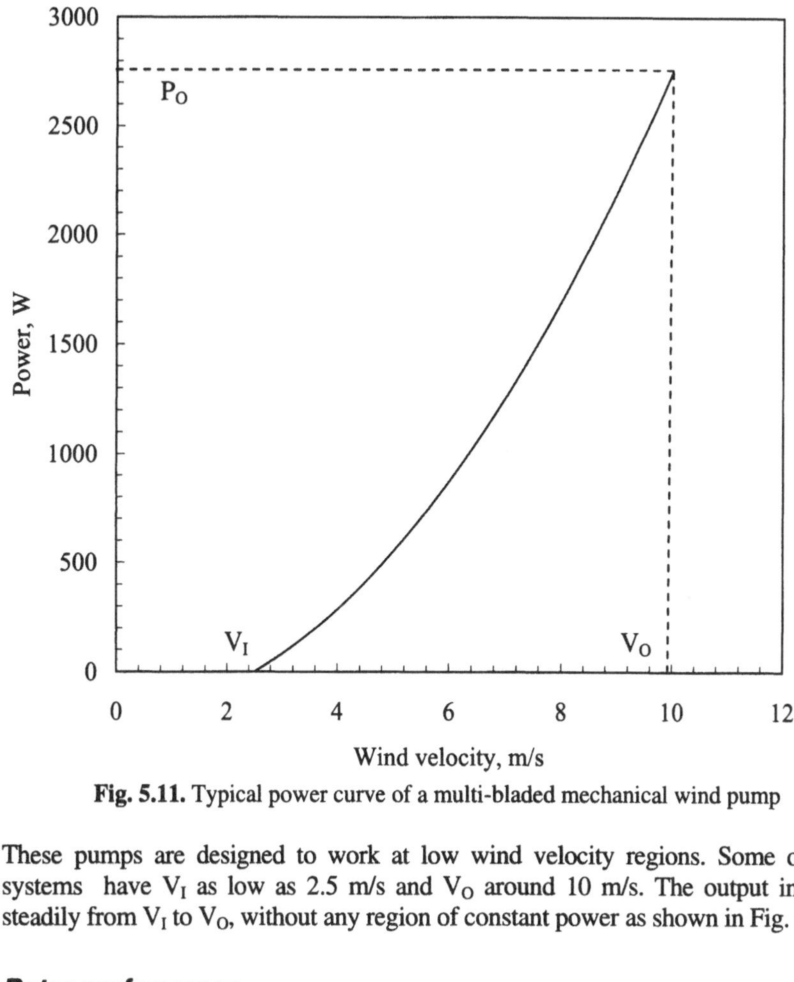

Fig. 5.11. Typical power curve of a multi-bladed mechanical wind pump

These pumps are designed to work at low wind velocity regions. Some of these systems have V_I as low as 2.5 m/s and V_O around 10 m/s. The output increases steadily from V_I to V_O, without any region of constant power as shown in Fig. 5.11.

Rotor performance

Usually, the rotors used for water pumping wind mills show quadratic velocity-power relationship. Hence, we can express the power P_V at any wind velocity V as

$$P_V = aV^2 + b \tag{5.30}$$

where a and b are constants. At the cut-in condition, the power is zero and thus

$$aV_I^2 + b = 0 \tag{5.31}$$

The system reaches at its maximum power at the cut-out point. Hence the power P_O, corresponding to V_O can be represented as

$$a\,V_O^2 + b = P_O \tag{5.32}$$

Solving Eqs. (5.31) and (5. 32) for a and b and substituting in Eq. (5.30), we get

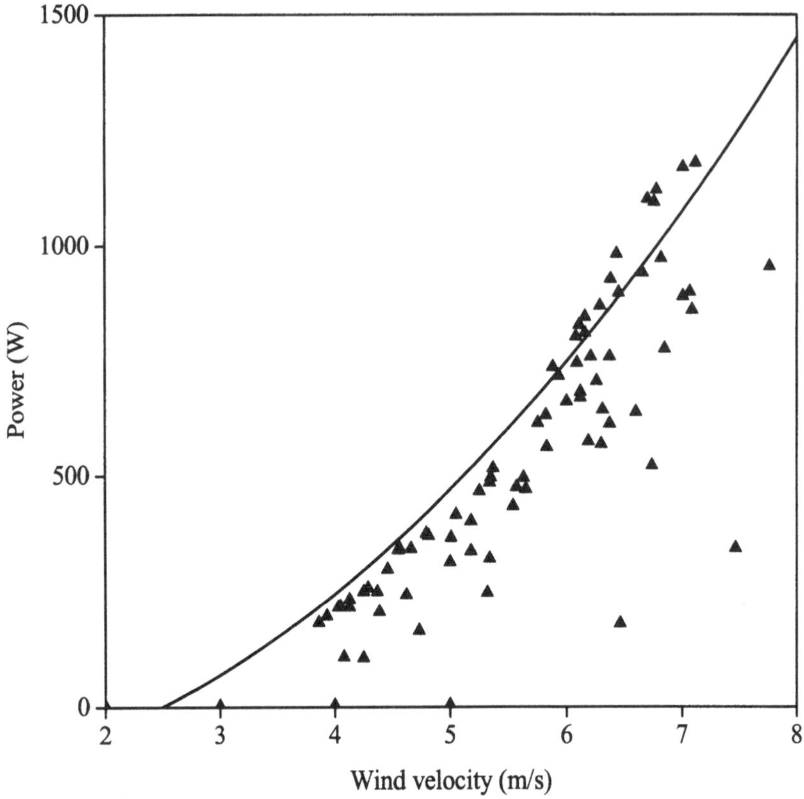

Fig. 5.12. Field performance of the experimental rotor

$$P_V = P_O\left(\frac{V^2 - V_I^2}{V_O^2 - V_I^2}\right) \tag{5.33}$$

The above equation is similar to the expression that we have developed for the wind generators. Test result of an eight bladed-5 m diameter experimental rotor developed for a water pumping wind mill are shown in Fig. 5.12. V_I and V_O of the rotor are 2.5 and 10 m/s respectively. Performance of the rotor generated using Eq. (5.33) is also indicated in the figure as continuous line. Reasonable agreement can be seen between the estimated and observed performances.

For estimating the energy produced by the rotor at a given site over a period, we have to introduce the wind regime characteristics in our analysis. The energy developed (E_I) over the period T is given by

$$E_I = T\int_{V_I}^{V_O} P_V\, f(V)\, dV \tag{5.34}$$

Assuming the Rayleigh distribution for wind velocity and substituting for f(V) and P(V), we get

$$E_I = T \int_{V_I}^{V_O} P_O \left(\frac{V^2 - V_I^2}{V_O^2 - V_I^2} \right) \frac{\pi}{2} \frac{V}{V_m^2} e^{-\pi/4 \left(V/V_m \right)^2} dV \tag{5.35}$$

Substituting

$$X = \frac{\pi}{4} \left(\frac{V}{V_m} \right)^2, \quad X_I = \frac{\pi}{4} \left(\frac{V_I}{V_m} \right)^2 \quad \text{and} \quad X_O = \frac{\pi}{4} \left(\frac{V_O}{V_m} \right)^2, \tag{5.36}$$

$$E_I = \frac{T P_O}{V_O^2 - V_I^2} \int_{X_I}^{X_O} \frac{4 V_m^2}{\pi} (X - X_I) e^{-X} dX \tag{5.37}$$

$$= \frac{T P_O}{V_O^2 - V_I^2} \frac{4 V_m^2}{\pi} \left\{ \left[-(X - X_I) e^{-X} - e^{-X} \right]_{X_I}^{X_O} \right\} \tag{5.38}$$

Further simplification yields

$$E_I = T P_O \left[\frac{4 V_m^2}{\pi \left(V_O^2 - V_I^2 \right)} \left(e^{-X_I} - e^{-X_O} \right) - e^{-X_O} \right] \tag{5.39}$$

Wind pump characteristics

The load imposed on the turbine by a positive displacement pump is typical in nature. For an unbalanced single acting pump, the starting torque is π times higher than the average running torque [2]. This means that, once the system stops due to the fluctuations in the wind regime, it restarts only at a wind speed strong enough to produce this high torque. Hence, while estimating the performance of such a wind pump, apart from the rotor and wind regime characteristics, the pump behaviour should also be taken into account. The overall performance coefficient of a wind rotor coupled to a reciprocating pump can be modelled as [3,9]

$$Cp\,\eta = 4\, Cp_d\, \eta_{(T,P)} \left[1 - K_O \left(\frac{V_I}{V} \right)^2 \right] K_O \left(\frac{V_I}{V} \right)^2 \tag{5.40}$$

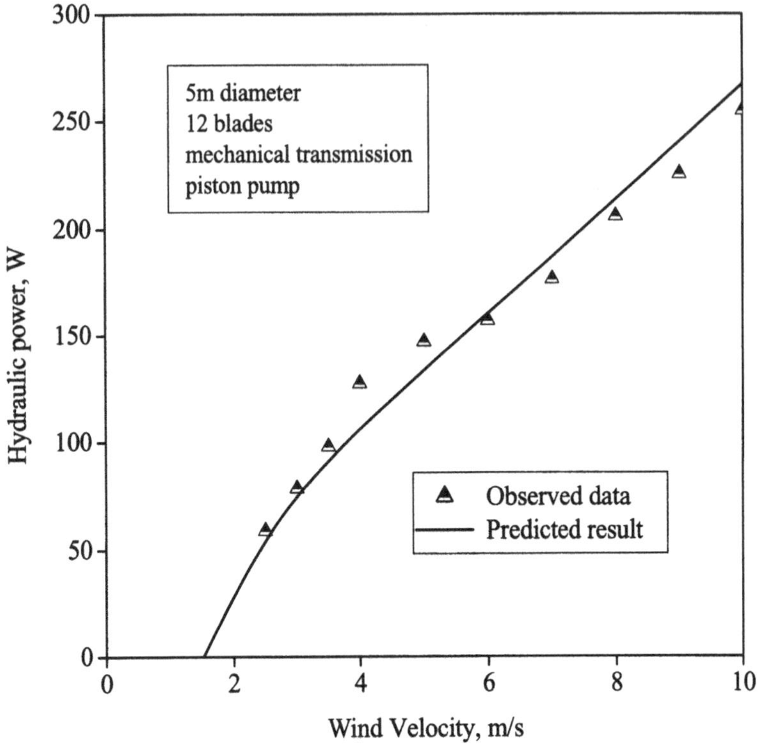

Fig. 5.13. Field performance of wind driven piston pumps

where $C_P\,\eta$ is the overall (wind to water) efficiency of the system, C_{pd} is the power coefficient of the rotor at the design point, $\eta_{(T,P)}$ is the combined transmission and pump efficiency and K_0 is a constant taking care of the starting behaviour of the rotor pump combination. For a single acting pump, K_0 varies from 0.2 to 0.25. The power developed by the system in pumping water (P_V) is given by

$$P_V = C_P\,\eta\,\frac{1}{2}\rho_a\,A\,V^3 \tag{5.41}$$

From Eqs. (5.40) and (5.41), P_V can be expressed as

$$P_V = 2\,C_{pd}\,\eta_{(T,P)}\,\rho_a\,A\,V^3\left[1-K_0\left(\frac{V_I}{V}\right)^2\right]K_0\left(\frac{V_I}{V}\right)^2 \tag{5.42}$$

Fig. 5.13 compares the field performances of a commercial wind pump, with the performance computed using Eq. (5.42). Close agreement can be observed between the measured and calculated performances.

The hydraulic power (P_H), needed by the pump for delivering a discharge of Q_{VP} against a head h is given by

$$P_H = \rho_w\,g\,Q_{VP}\,h \tag{5.43}$$

Equating the power delivered by the rotor and absorbed by the pump, and solving for Q_{VP}, the instantaneous discharge of the system at any velocity V can be derived. Thus,

$$Q_{VP} = 2\,Cp_d\,\eta_{(T,P)}\left[\frac{\rho_a}{\rho_w}\right]\left[\frac{A\,V^3}{gh}\right]\left[1-K_0\left(\frac{V_I}{V}\right)^2\right]K_0\left(\frac{V_I}{V}\right)^2 \tag{5.44}$$

Integrated system performance

Inorder to estimate the discharge of the wind pump over a period, let us integrate the rotor, pump and wind regime characteristics into our model. In light of Eq. (5.42), the power P_O corresponding to the velocity V_O can be rewritten as

$$P_o = 2\,\rho_a\,A V_O^3\;\;Cp_d\,\eta_{(T,P)}\left[1-K_0\left(\frac{V_I}{V_O}\right)^2\right]K_0\left(\frac{V_I}{V_O}\right)^2 \tag{5.45}$$

Similarly, the energy required to pump a total discharge of Q_{IP} over a period is given as

$$E_H = \rho_w\,g\,Q_{IP}\,h \tag{5.46}$$

Substituting for P_O in the expression for E_I in Eq. (5.39), equating it with hydraulic energy demand E_H, and solving for Q_{IP}, we get

$$Q_{IP} = 2\,T\,CP_d\,\eta_{(T,P)}\,\frac{\rho_a}{\rho_w}\,\frac{A_T V_O^3}{g\,h}\left[1-K_0\left(\frac{V_I}{V_O}\right)^2\right]K_0\left(\frac{V_I}{V_O}\right)^2 \tag{5.47}$$

$$\left[\left\{\frac{4\,V_m^{\;2}}{\pi\left(V_O^2-V_I^2\right)}\left(e^{-X_I}-e^{-X_O}\right)\right\}-\left\{e^{-X_O}\right\}\right]$$

Thus Eq. 5.47 gives us the discharge expected from a wind driven piston pump, installed at a given site, over a period T.

Example

Generate the velocity-discharge characteristic of a wind driven piston pump working against a head of 5 m. Specifications of the pump are given in Table.5.8.

Let us take ρ_a and ρ_w as 1.24 kg/m^3 and 1000 kg/m^3 respectively. As diameter of the rotor is 3 m, its area is 7.065 m^2. Substituting the system specifications in Eq. (5.44), we get

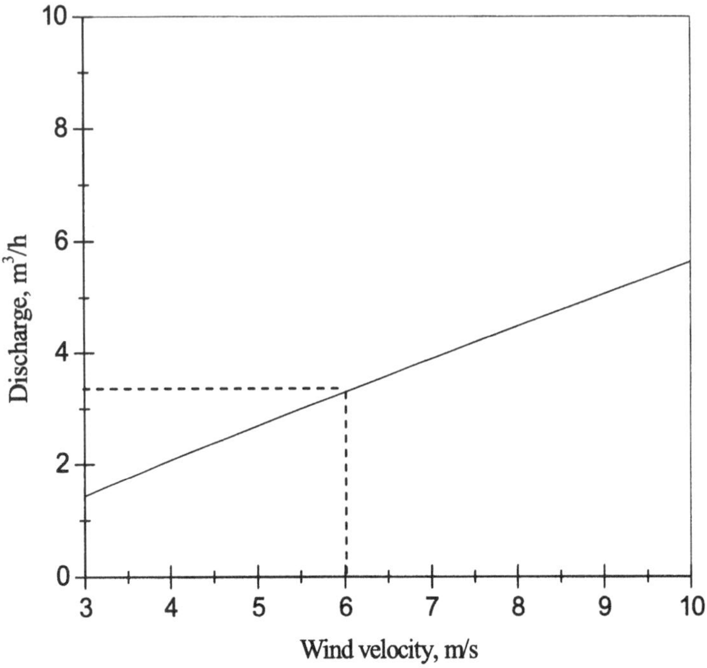

Fig. 5.14. Performance of the wind driven piston pump

Table 5.8. Specifications of the wind driven piston pump

Diameter of the rotor, m	3
Combined efficiency of pump and transmission system	0.95
Coefficient K_0	0.25
Cut – in velocity, m/sec	2.5

$$Q_{VP} = \frac{1}{2}\left[\frac{1.24}{1000}\right]\left[\frac{7.065\ V^3}{9.8\times 5}\right]\left[\begin{array}{c}4\times 0.3\times 0.95\times \\ \left[1-0.25\times\left(\frac{2.5}{V}\right)^2\right]\times 0.25\times\left(\frac{2.5}{V}\right)^2\end{array}\right]$$

By calculating Q_{VP} for different V, we can develop the performance curve of the wind pump as shown in Fig. 5.14.

Example

Monthly average wind velocities at a site are given in Table 5.9. Calculate the discharge of the wind pump in the above example, at this site. Pumping head is 5.

Table 5.9. Monthly average wind velocity (m/s)

Jan	Feb	Mar	Apr	May	June	July	Aug	Sept	Oct	Nov	Dec
6.21	5.40	4.83	4.47	4.71	4.73	4.49	4.61	3.74	3.92	4.13	4.89

Let us solve this problem using the "WIND PUMP-POSITIVE DISPLACEMENT" module of the WERA programme. The results are presented in Table. 5.10.

Table 5.10. Monthly discharge of the wind pump

Month	V_m, m/s	Discharge , m^3
January	6.21	1030.79
February	5.4	943.95
March	4.83	830.22
April	4.47	736.02
May	4.71	800.61
June	4.73	805.67
July	4.49	741.66
August	4.61	774.54
September	3.74	508.56
October	3.92	567.23
November	4.13	634.26
December	4.89	844.31

5.5.2 Wind driven roto-dynamic pumps

Wind turbines mechanically coupled with roto-dynamic pumps are increasingly getting attention in the recent years. These pumps are suitable for pumping large quantities of water from shallow wells and ponds. Let us consider the performance of such a system under fluctuating conditions of wind regimes. Here we assume that the efficiency of the combined system peaks at the operating point where the efficiency of the pump is also at its maximum.

The power (P_{Td}) developed by a wind turbine at its design wind speed (V_d) is given by

$$P_{Td} = \frac{1}{2} C_{Pd} \ \rho_a \left[\frac{\pi \ D_T^{\ 2}}{4} \right] V_d^{\ 3} \tag{5.48}$$

Where C_{Pd} is the design power coefficient and D_T is the rotor diameter. Considering the characteristics of an ideal roto-dynamic pump, the power required (P_{pd}) for delivering its design discharge (Q_d) against the design head (h_d) may be expressed as

$$P_{Pd} = \frac{\rho_w g \ h_d \ Q_d}{\eta_{pd}} \tag{5.49}$$

where η_{pd} is the design efficiency of the pump.

Assuming that both the rotor and the pump are matched at their peak efficiency points, we can equate the power developed by the turbine with that required by the pump. Thus, discharge of the combined system at its design point can be solved as

$$Q_d = \frac{1}{2} C_{pd}\ \eta_{pd}\ \eta_T\ \frac{\rho_a}{\rho_w}\left[\frac{\pi\ D_T^{\ 2}}{4}\right]\frac{V_d^{\ 3}}{g\,h_d} \tag{5.50}$$

Here η_T is the efficiency of power transmission from the wind rotor to the pump. As the wind pump is exposed to fluctuating conditions of wind regime, the off design performance of the system is also important. For an ideal roto-dynamic pump, the discharge (Q), pump speed (N_P) and the pump diameter (D_p) can be correlated as

$$Q\ \alpha\ N_p\ D_p^{\ 3} \tag{5.51}$$

Thus, the discharge of the wind pump at any velocity V, may be expressed in terms of its design discharge as

$$Q_V = Q_d\left(\frac{N_{pv}}{N_{pd}}\right) \tag{5.52}$$

where Q_V and N_{pv} are the discharge and pump speed at velocity V and N_{pd} is the design pump speed.

As the pump, at all its operating speeds, is matched with the best efficiency points of the turbine, the rotational speed of the pump at any velocity can be represented by

$$N_{pv} = G\ \lambda_d\ V\left(\frac{1}{\pi\ D_T}\right) \tag{5.53}$$

where G is the gear ratio to be introduced between the wind rotor and pump and λ_d is the design tip speed ratio of the rotor. From Eqs. (5.50), (5.52) and (5.53), the pump discharge at any wind velocity can be expressed as

$$Q_V = \frac{1}{8} C_{pd}\ \eta_{pd}\eta_T\ V\ D_T\left(\frac{\rho_a}{\rho_w}\right)\left(\frac{V_d^{\ 3}}{g\ h_d}\right)\left(\frac{G\ \lambda_d}{N_{pd}}\right) \tag{5.54}$$

Eq. (5.54) gives us the system discharge against the design head. However, due to changing field conditions, the lift requirement also may vary. As for the same power discharge is inversely proportional to head, we have

$$Q_{(V,h)} = Q_V\ \frac{h_d}{h} \tag{5.55}$$

where $Q_{(V,h)}$ is the discharge of the system at a velocity V against a head h. Combining Eqs. (5.54) and (5.55), we can express $Q_{(V,h)}$ as

$$Q_{(V,h)} = \frac{1}{8} C_{pd} \ \eta_{pd} \eta_T \ V \ D_T \left(\frac{\rho_a}{\rho_w} \right) \left(\frac{V_d^{\ 3}}{g \ h} \right) \left(\frac{G \ \lambda_d}{N_{pd}} \right) \tag{5.56}$$

If N_{pi} be the rotational speed of the pump required to start pumping against a given head, then

$$N_T = \frac{N_{pi}}{G} \tag{5.57}$$

Hence, the cut-in velocity of the wind driven roto-dynamic pump at a certain hydraulic load can be expressed by

$$V_I = \frac{\pi \ D_T \ N_{pi}}{G \ \lambda_d} \tag{5.58}$$

Integration of wind regime characteristics

Let us now introduce the wind regime characteristics in the analysis to quantify the pump discharge, at a given site, over a period. Here we assume that the system cuts in at V_I and cuts out at V_O without any region of constant power. With this assumption, discharge of the wind pump over a period (Q_{IR}) is given by [10]

$$Q_{IR} = T \int_{V_I}^{V_O} Q_{(V,h)} \ f(V) \ dV \tag{5.59}$$

Assuming the Rayleigh distribution for the wind velocity and with

$$X = \frac{\pi}{4} \left(\frac{V}{V_m} \right)^2 \tag{5.60}$$

Q_{IR} can be represented as

$$Q_{IR} = \frac{\pi}{16} \ T D_T \ C_{pd} \ \eta_T \eta_{pd} \ \left(\frac{\rho_a}{\rho_w} \right) \left(\frac{V_d^{\ 3}}{V_m^{\ 2}} \right) \left(\frac{1}{g \ h} \right) \left(\frac{G \ \lambda_d}{N_{pd}} \right) \int_{V_I}^{V_O} V^2 \ e^{-X} \tag{5.61}$$

Example

Specifications of a wind powered roto-dynamic pump are given in Table 5.11. Generate the performance curve of the system at 5 m head and locate the pump discharge at 6 m/s wind velocity.

Table 5.11. Specifications of the wind driven roto-dynamic pump

Specifications	
Diameter of the rotor, m	3.00
Design power coefficient	0.38
Efficiency of the pump	0.62
Transmission efficiency	0.74
Design wind velocity, m/s	5.00
Gear ratio	16.00
Design tip-speed ratio	2.50
Pump speed at design point, r/s	23.00

For generating the performance curve, we have to establish a relationship between the wind velocity and corresponding pump discharge at 5 m head. Inserting the system specifications in Eq. (5.56), we get

$$Q(V,h) = \left(\frac{1}{8} \times 0.38 \times 0.62 \times 0.74 \times 3 \times \left(\frac{1.24}{1000}\right) \left(\frac{5^3}{9.81 \times 3}\right) \times \left(\frac{16 \times 2.5}{23}\right)\right)$$

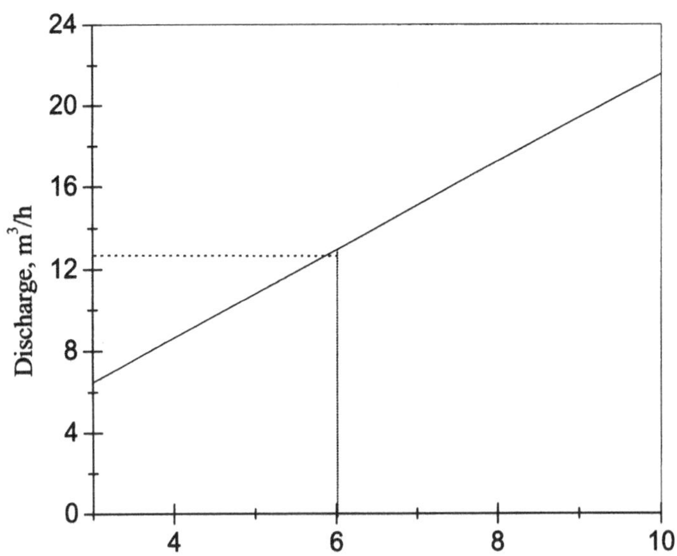

Fig. 5.15. Performance curve for the wind driven roto-dynamic pump

Plotting this relationship between $Q_{(V,\ h)}$ and V, we get the performance curve of the wind pump as shown in Fig. 5.15. From the figure, we can see that at 6 m/s wind velocity, the system delivers 12.93 m^3/h against 5 m pumping head.

Example

Consider the monthly average wind velocities given in Table 5.9. A wind driven roto-dynamic pump with specifications as in Table 5.11 is installed at this site. Estimate the monthly output from the pump against 5 m head.

Let us use the "WIND PUMP-ROTODYNAMIC" module of the WERA programme to solve this problem. Estimated wind pump discharges for different months are given in Table. 5.12

Table 5.12. Monthly performance of the wind driven roto-dynamic pump

Month	V_m, m/s	Discharge, m^3
January	6.21	4020.56
February	5.40	3958.06
March	4.83	3757.94
April	4.47	3554.18
May	4.71	3696.82
June	4.73	3707.47
July	4.49	3567.10
August	4.61	3640.68
September	3.74	2963.30
October	3.92	3129.02
November	4.13	3306.68
December	4.89	3785.96

5.5.3 Wind electric pumping systems

Wind pumps with electrical transmission are more efficient, reliable and flexible compared to the mechanical systems. These pumps have the performance regions similar to that of the wind electric generators described in section 5.1. Detailed performance models for wind electric pumps with variable speed permanent magnet generators and induction motor drives are discussed in [11]. Here, we adopt a simpler method to approximate the performance of these pumps.

Referring to Fig. 5.1, in the first performance region we have

$$\frac{\rho_{w}\, g\, h\, Q}{\eta_p} = P_R \left(\frac{V^n - V_I^n}{V_R^n - V_I^n} \right) \tag{5.62}$$

From the above expression, discharge of the pump (Q_V) at velocity V and head h can be solved as

$$Q_V = \frac{\eta_p}{\rho_w\, g\, h} P_R \left(\frac{V^n - V_I^n}{V_R^n - V_I^n} \right) \tag{5.63}$$

where η_p is the pump efficiency. For the second region of constant power,

$$Q_V = \frac{\eta_p}{\rho_w\, g\, h} P_R \tag{5.64}$$

To estimate the integrated discharge from these pumps over a period, the energy developed by the system may be computed as discussed in case of the wind generators and then equated with the corresponding hydraulic energy demand. Thus

$$Q_{IE} = \frac{\eta_p\, E_I}{\rho_w\, g\, h} \tag{5.65}$$

Example

Specifications of a wind electric pump are given in Table 5.13. Generate the performance of the pump at different wind velocities.

Table 5.13. Specifications of the wind electric pump

Rated power	1400 W
Cut-in wind speed	3 m/s
Rated wind speed	10 m/s
Cut-out wind speed	15 m/s
Efficiency (pump+ power conversion)	0.52
Pumping head	5 m
Velocity power proportionality	2

Substituting the system specifications in Eq. (5.63), we get

$$Q_V = \frac{0.52}{1000 \times 9.81 \times 5} \times 1400 \times \left(\frac{V^2 - 3^2}{10^2 - 3^2} \right)$$

Thus we have the relationship between the wind velocity and discharge in the 1st performance region. In the 2nd performance region, discharge is constant, which corresponds to the rated wind velocity. The generated performance curve of the pump is shown in Fig. 5.16.

In the above sections, we have discussed the performance models for three different types of wind pumps.

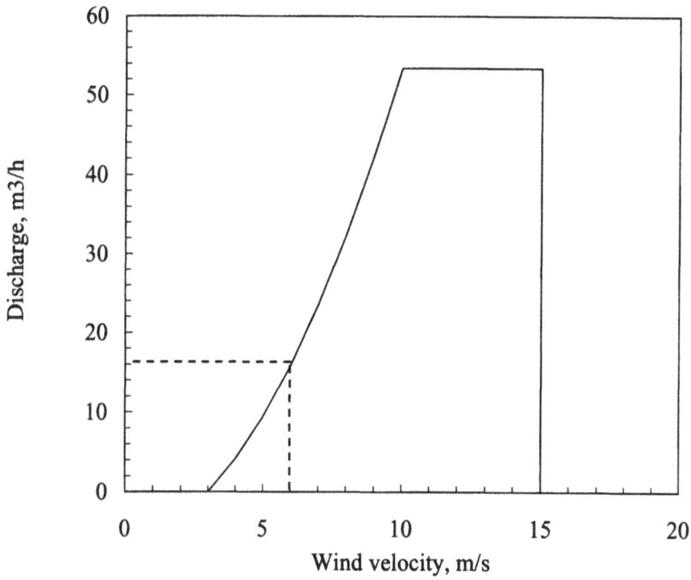

Fig. 5.16. Performance curve for the wind electric pump

It is interesting to compare the performances of these pumps under similar conditions. Let us consider a wind speed of 6 m/s and pumping head of 5 m. We can see that the discharge of the wind driven piston pump under these conditions is 3.3 m^3/hr (Fig. 5.14), whereas the corresponding rates of discharge for roto-dynamic and wind electric pumps are 12.9 m^3/h and 15.8 m^3/h respectively (Figs. 5.15 and 5.16). The efficient performance of the latter systems are due to the better matching between the wind rotor and the roto-dynamic pumps coupled to it, as we have seen in Chapter 4.

References

1. Biswas S, Sreedhar BN, Singh YP (1994) A simplified statistical technique for wind energy output estimation. Wind Eng. 15 (10) : 921-930
2. Burton JD (1983) Double acting pump with inertia flow improves the load matching of water pumping windmills. In: Seventh Conference of Fluid Machinery, Budapest
3. Burton JD, Pinilla AE (1985) Water pump for wind mills – A comparison between two commercially available systems from South America. Wind Eng. 9 (1) : 50-58
4. Galanis N, Christophides C (1990) Technical and economical considerations for the design of optimum wind energy conversion systems. J Wind Eng. Ind. Aerodynamics 34 : 185-196
5. Jangamshetti SH, Rau VG (2001) Optimum siting of wind turbine generators. IEEE Transactions on Energy Conservation 16 (1) : 8-13
6. Johnson GL (2001) Wind energy systems. http://www.rpc.com.au

7. Li S, Wunsch DC, O'Hair EA, Giesselmann MG (2001) Using neural networks to estimate wind turbine power generation. IEEE Transactions on Energy Conversion 16 (3) : 276-282
8. Lysen, E. H.,1983. Introduction to Wind Energy. CWD, Amersfoort, The Netherlands
9. Mathew S, Pandey KP (2000) Modelling the integrated output of mechanical wind pumps. J Solar Energy Engineering 122
10. Mathew S, Pandey KP (2003) Modelling the integrated output of wind-driven roto-dynamic pumps. Renewable Energy 28 : 1143-1155
11. Mathew S, Pandey KP, Burton JD (2002) The wind-driven regenerative water-pump. Wind engineering 26(5) : 301-313
12. Mathew S, Pandey KP, Kumar AV (2002) Analysis of wind regimes for energy estimation. Renewable Energy 25: 381-399
13. Pallabazzer R (1995) Evaluation of wind – generator potentiality. Solar Energy 55 (1) : 49-59
14. Powel WR (1981) An analytical expression for the average output power of a wind machine. Sol Energy 26 : 77-80

6. Wind energy and environment

Fossil fuel based power plants, contributing more than seventy per cent to our energy needs today, dominate the global energy scenario. These plants pollute the atmosphere with harmful gasses and particulates. As per the estimates of the International Energy Agency (IEA), 23683 Mt of CO_2 has been released to the atmosphere by the power sector during 2001 [15]. Emission of CO_2 due to power generation has registered an increase of 65 per cent in during the past three decades. With the increase in energy demand, level of environmental pollution caused by the power sector is expected to increase further in the coming years.

In contrast, wind energy does not pollute the air or water with harmful gases and materials. Nor does it generate hazardous wastes, which cannot be safely disposed as in case of nuclear power plants. Being a non depletable source, extracting energy from wind does not pose the threat of over exploiting the limited natural resources like coal, oil or natural gas. Hence wind is considered as one of the cleanest sources of energy available today.

If we can exploit even a small fraction of this abundant and environment-friendly source of energy, today's power related emissions can be reduced to a much acceptable level. For example, if 10 per cent of the wind potential in US is effectively exploited, the total carbon dioxide emission from the country can be reduced by 33 per cent, which is equivalent to a reduction of 4 per cent in the global level. Similarly, by meeting 20 per cent of the countries energy demand from wind, the emission from coal fired power plants can be reduced by one third [5].

As in case of any human activities, wind energy generation is also not totally free from environmental consequences. The major environmental problem with wind energy is avian mortality due to collision with turbines and related structures. Noise emission and the visual impacts on landscapes are the other issues to be tackled. However, it should be noted that these environmental impacts are not global (as in case of atmospheric emissions and global warming) and thus can be monitored and resolved at local level.

In this chapter, let us examine the merits and demerits of wind energy in an environmental perspective. We will adopt a 'life cycle' approach in our discussions. This means that, while dealing with the atmospheric emissions, apart from the pollution due to power generation, possible emissions during the turbine manufacturing, plant construction etc. will also be accounted. The major environmental ill effects of wind energy are then considered with discussions on possible solutions.

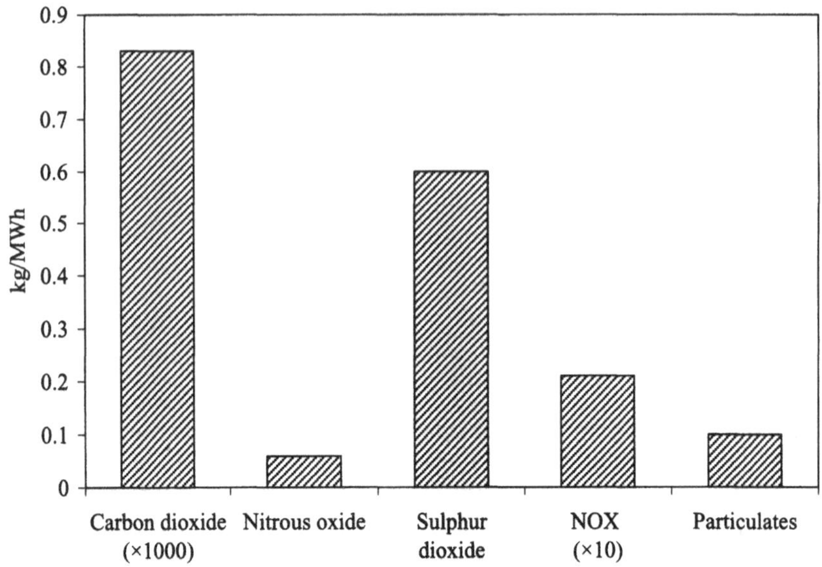

Fig. 6.1. Atmospheric emissions from a coal based power plant

6.1 Environmental benefits of wind energy

The most significant environmental advantage of wind energy is that it does not load the atmosphere with toxic chemicals. In contrast, the conventional power plants operating on fossil fuels release gases like sulphur dioxide, carbon dioxide and oxides of nitrogen during the energy conversion process. Similarly, particulates and toxic heavy metals which affect the environment adversely are also being generated during this process.

For example, the specific emission rates from a typical power plant based on pulverized coal is shown in Fig. 6.1[1]. The fuel used here is bituminous coal and the flue gases are undergone desulphurization using the lime stone/gypsum. It should be noted that ninety per cent of the sulfur dioxide are removed under this process and still the level of emission is higher than the acceptable limit.

Emission due to the fossil fuel based generation depends on the type and quality of fuel used and the technology of power conversion. For example, in case of a modern natural gas-combined cycle power plant, the level of SO_X emission is relatively low. Here, the major problems are due to NO and NO_2 emissions. Typical concentration of CO_2 in such plants may be nearly 10 ppmv (parts per million by volume) where as N_2O levels may be around 1 ppmv. Even with advanced technologies like the Integrated Gasification Combined Cycle (IGCC), the CO_2 level could be as high as 794 kg /MWh.

Sulphur dioxide and nitrogen oxides released to the atmosphere due to the burning of fossil fuels cause acid rain. When SO_2 and NO_x combine with water in the atmosphere, a mixture of sulfuric and nitric acid is formed. These acids are then washed out of the atmosphere by rain and thus reach forests, lakes and streams. Acid rain is one of the most dangerous forms of pollution. Presence of acid rain in an area may not be noticed for several years. It creates severe health and environmental impacts. Acidification of lakes and streams kills fish and amphibians. Increase in the acidity of water will also reduce the reproductive capacity of aquatic organisms. As a result, the lakes and streams gradually become 'dead'. In turn, this will be harmful to other species of birds and mammals which depend on aquatic organisms for their food. Many lakes in US have become biologically dead due to this form of pollution.

Acid rain also damages the forests as it strips the essential nutrients off the soil, which in turn increases the concentration of toxic metals. It also damages the foliage of trees and retards their growth. Many forests in Poland and Czechoslovakia are already damaged-to an extent of 40 per cent-due to acid rain. Human beings are also affected by the presence of toxic acids in the atmosphere. Defensive mechanism of lungs are affected which causes respiratory problems like bronchitis and asthma. If the drinking water is contaminated due to this form of pollution, then the human beings are widely harmed. Acid rain causes damage to buildings and structures as well.

Three quarter of the global emission of CO_2 is due to the burning of fossil fuels. Accumulation of carbon dioxide in the atmosphere causes the green house effect and thus the global warming. The green house gases permits the sun's rays to enter the atmosphere but traps the reflected infrared radiation. This causes the atmosphere to get warmer. The global warming causes the weather pattern to change, the sea level to rise and the land use configuration to alter.

Another problem of fossil fuel based power generation is the formation of smog. Smog is a combination of smoke and fog. It is mainly constituted with the ground level ozone, which is produced by the reaction between the nitrous oxides and volatile organic compounds. NO_X is mainly generated due to the burring of fossil fuels. Smog is a severely irritating pollutant. It can affect nose and throat and cause respiratory problems such as severe cough and deep breathing. Over 1800 people are reported to be killed by fog every year in Ontario, Canada [12]. This can also be hazardous to crops and other vegetations. The losses in agricultural production due to smog in Ontario region are estimated to be $70 million. Similarly, particulates and toxic heavy metals present in the atmosphere causes asthma and lung cancer.

The fossil fuel based plants consume a lot of water during the thermodynamic cycle. For instance, the typical evaporative losses in nuclear plant may be as high as 2.3 l/kWh. Losses in the coal and oil based plants are also high (approximately 1.9 l/kWh and 1.6 l/kWh). Considerable quantity of water is consumed in cleaning the fuel in case of the coal fired plants. This may be a severe problem in areas where water is scarce. In contrast, water requirement for wind energy plants are negligible. We require water only for the periodic cleaning of the blades.

Nuclear energy-which is often projected as the energy source for the future by some corners-is also not free from environmental risks. All the processes involved in the nuclear energy technology are prone to environmental hazards. Radioactive isotopes, which are generated right from the mining of uranium, its processing and enrichment and plutonium production, may pollute the ground water, land and plants. These may be active for several thousands of years, posing threat to the entire ecosystem. Further, a safe and reliable way to dispose the hazardous waste produced at different stages of the nuclear energy generation is yet to be discovered. Another problem with this technology is the high risk of accidents. We still remember the unfortunate accident at the Chernobyl plant (presently in Ukraine) in 1986. The level of radiation, even at far away places like Scotland (about 2300 km away from the accident spot), were 10,000 times the norm. Thousands of people has been killed due to cancer and other diseases as the result of this disaster.

Here comes the significance of the clean sources like wind in meeting our energy demand in a sustainable and environment-friendly way. Wind energy systems neither generate polluting gasses, nor does it load the environment with harmful particulates. The technology also does not pose problems of radiation and waste management.

Example

Estimate the mitigation potential of a wind farm with 5 turbines of 1 MW rated capacity. The cut-in, rated and cut-out velocities of the turbines are 3.5, 12.5 and 25 m/s respectively. The Weibull shape and scale factors at the wind farm location are 2.45 and 8.2 m/s respectively.

Energy generated from the farm is calculated using the WERA programme using its WIND TURBINE-WEIBULL module. With the given turbine specifications and ideal cubic velocity-power relationship, the capacity factor of the system at this site is found to be 0.273. Thus the total energy generated from the farm is

$$0.273 \times 1000 \times 8760 \times 5 = 11957400 \text{ kWh/year}$$

Assuming 90 per cent efficiency for grid integration, the total annual energy out put of the farm is 10761.66 MWh. Now, comparing with the coal based generation (Fig. 6.1), each MWh generated from the farm would mitigate 830 kg of CO_2, 2.1 kg of NO_X, 0.6 kg of SO_2, 0.1 kg of particulates and 0.06 kg of Nitrous oxide. Based on these emission levels, the mitigation potential of the wind farm is calculated as shown in Table 6.1.

6.2 Life cycle analysis

In the above example, we have seen that the wind farm could mitigate an impressive amount of atmospheric emissions. A few questions that may arise at this point are (1) Don't we require energy for manufacturing the turbine and constructing the plant? (2) Shouldn't we account the emissions in the manufacturing and commissioning stages of the system in our analysis? (3) Shouldn't we consider the energy required for disposing the plant after its life period? (4) If we consider these factors, can we say that the process of generating energy from wind is totally free from atmospheric emissions? (5) How rapidly could the system recover all the energy consumed for its manufacturing, installation, operation and dismantling?

Table 6.1. Annual reduction in atmospheric emissions due to a 5 MW wind farm

Pollutant	Reduction, kg
Carbon dioxide	8932178
Nitrous oxide	645.6996
Sulphur dioxide	6456.996
NO_X	22599.49
Particulates	1076.166

To answer these questions, we should broaden our analysis to the entire life period of the project. This is called the life cycle analysis (LSA).

In the life cycle based analysis, we look at a system or technology in its totality and account the energy use and related emissions involved in all the stages of its production, use and disposal. This essentially should include the extraction of raw materials, its conversion into different components, manufacturing, commissioning and operation of the system, and finally its disposal or recycling after use. For example, the life cycle of a typical wind turbine is shown in Fig. 6.2. When we assess such systems, it is logical to account the energy flow and emission potential of all the phases of the project life as indicated in the figure. Here, upto the phase of turbine manufacturing, energy of various forms are consumed and thus the system will have a negative impact on environment. In contrast, during the operational phase, energy is being generated without any pollution and thus the project reacts positively to the environment.

As all the processes involved in the project-right from the inception to the decommissioning- are considered, the LSA is often termed as a 'cradle to grave' method. LSA is a versatile approach for assessing and comparing the environmental merits of different energy sources and technologies. Further it can be used as an effective tool for identifying possible improvements in the system design and decision making.

The LSA broadly consists of five stages as shown in Fig. 6.3. In the first stage, the goal or the scope of the analysis is defined. The unit processes involved in the life cycle of the technology (production, use and disposal) are identified at this stage. The next step is to clearly lay down the assumptions which form the basis of our analysis. Further, in the inventory analysis stage, the inflow and outflow of relevant materials and energy, throughout the life cycle of the system, is quantified on the basis of reliable data. In the impact assessment phase, the environmental implications of the material and energy flow are analyzed. For easy comparison, impacts of different kinds are often weighed and normalized on the basis of their intensity of environmental loading. Finally, during the interpretation and design improvement phase, results of the previous stages are analyzed and possible design improvements in the system are identified with a detailed plan of action.

The first step in the LSA of a wind turbine is to quantify the material required for the system. This should include the materials used for the fabrication of different turbine components and the foundation. Type and quantity of materials required for a system varies with its size (kW rating), design features and site conditions. Approximate amount of materials required for a typical 600 kW land based turbine is given in table 6.2. As offshore turbines require stronger foundation and

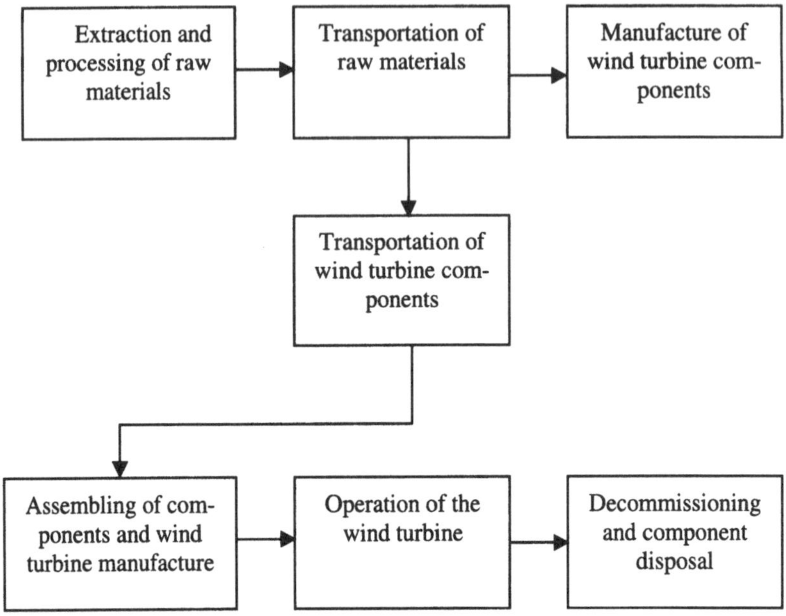

Fig. 6.2. Various stages of the life cycle of a wind turbine

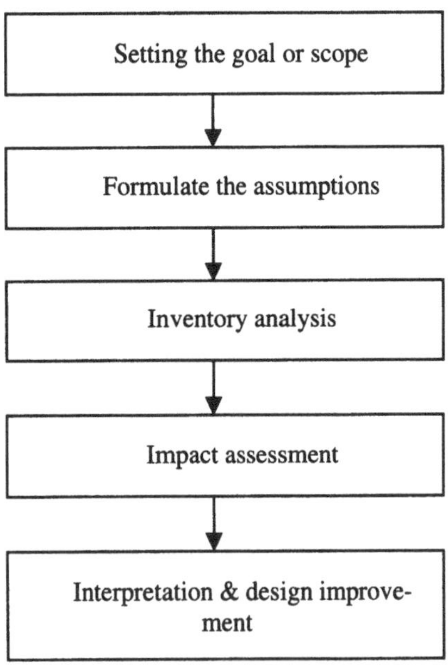

Fig. 6.3. Various stages involved in LSA

sea cabling, additional materials are required for such systems. Materials additional required for a 600 kW offshore turbine are listed in Table 6.3.

Table 6.2. Materials required for a 600 kW land based wind turbine

Materials	kg
Steel	63460
Aluminum	1780
Copper	740
Sand	2520
Glass	1320
Polyester and epoxy	2400
Oil products	120
Reinforced iron (foundation)	14400
Concrete (foundation)	339000
Miscellaneous	840

Table 6.3. Additional materials required for the 600 kW offshore turbine

Materials	kg
Reinforced iron	14400
Concrete	339000
Copper	3096
Lead	4032
Steel	4680
PEX	648

Reliability of the LSA greatly depends on the accuracy in estimating the materials used for the system. Estimate for turbines of specific size and design can be obtained from the manufacturers on request.

There are two common ways to identify the environmental merit of an energy system under LSA. They are (1) Net Energy Analysis (NEA) and (2) Life Cycle Emission (LCE) analysis.

6.2.1 Net energy analysis

In the NEA, we compare the useful energy produced by the system (E_P) with the energy consumed by it throughout its life cycle (E_{CL}). The ratio between the energy developed and consumed by the system is often termed as the Energy Payback Ratio (EPR). Thus,

$$EPR = \frac{E_p}{E_{CL}} = \frac{E_A\ L}{E_{CL}} \tag{6.1}$$

where E_A is the annual energy production from the system and L is the life period. All the energy consumed during the manufacturing, operation and disposal phases of the technology should be included in E_{CL}. Another point of interest is to estimate the time required for the system to pay back all the energy consumed by it. This is called the Energy Payback Period (EPP). Thus we have

$$EPP = \frac{E_{CL}}{E_A} \tag{6.2}$$

The energy produced by the turbine at a given location is basically a function of its capacity factor. This means that, apart from the turbine type and size, the site characteristics also influence EPR and EPP. Even with a moderate capacity factor, wind farm projects generally pay back the energy consumed throughout its life cycle within one year of its commissioning. Modern wind turbines are more efficient and use fewer materials and hence pay back the energy much quicker than the earlier designs.

For EPR and EPP calculations, we require the amount of energy spent in manufacturing different components of the turbine. Once the quantity of different materials required for the components are identified (as in Table 6.2 and 6.3), then the energy input in respect of these components can be calculated by multiplying the weight of the materials used by the specific energy required to produce them. The specific energy depends on the energy consumed from various sources in producing these materials. For example, if the energy consumption in producing one kg of steel is 1.6 MJ equivalent of coke, 17.4 MJ equivalent of coal, 6.6 MJ equivalent of oil and 0.1 MJ equivalent of natural gas, then the specific energy of steel is 25.7 MJ/kg.

Table 6. 4. Energy equivalent of different materials used in a 600 kW land based turbine

Materials	Energy, MJ
Steel	1630922
Aluminum	69776
Copper	57868
Sand	9273.6
Glass	12276
Polyester and epoxy	109680
Reinforced iron (foundation)	522720
Concrete (foundation)	1247520
Miscellaneous	39228

Table 6.5. Energy requirement of additional materials used in a 600 kW offshore turbine

Materials	Energy, MJ
Reinforced iron	522720
Concrete	1247520
Copper	242107.2
Lead	143539.2
Steel	120276

Energy use pattern of a 600 kW wind turbine is given in Table 6.4 and 6.5. This includes the energy associated with the production, transportation and manufacture of the materials. The above calculations are based on the Danish conditions [9,28], where coal is the predominant energy source. The energy required to produce the materials may differ from place to place, depending on the locally available energy source. When we have a different situation-for example, in a country where hydro or nuclear power are the major sources-the scenario will change accordingly. The technology adopted for producing the materials is another factor influencing the energy use. For example, energy equivalent of steel can vary from 20.7 MJ/kg to 30.6 MJ/kg, based on the nature of the processes involved in its production. Thus, it should be ensured that the values assumed in our calculations are valid for the specific region of interest.

From the above tables, it is seen that the total energy consumed by a land based wind farm, in respect of its material of construction is 3,699.6 GJ. For an offshore system, the corresponding energy requirement is 5,975.5 GJ. Wind turbines do not consume energy during the operational phase of its life cycle. However, energy is required for decommissioning the system after its life period as we require energy for dismantling, transporting, and disposal or recycling the components. The energy demand in this regard can be taken as 0.1-0.4 MJ/kg [28] of the materials used. Thus, the energy required for disposing the 600 kW land based turbine is 106.6 GJ where as that for the offshore units is 198.1 GJ. This makes the total energy requirement of the land based turbine, during its life cycle as 3806.2 GJ. For the offshore system, corresponding energy requirement is 6173.6 GJ.

Example

A land based wind farm consists of 15 wind turbines of 600 kW size. The cut-in, rated and cut-out velocities of the turbines are 3.5, 12.5 and 25 m/s respectively. The Weibull shape and scale factors at the farm site are 2.8 and 8.2 m/s. Calculate the Energy Payback Ratio and Energy Payback Period of the farm.

From the previous section, we have seen that the total energy requirement of a 600 kW land based wind turbine is 3806.2 GJ. Hence, the total energy input for the farm with 15 turbines is 57093 GJ, which is 15859 MWh.

Assume that the velocity-power relationship of the turbines is quadratic. With the Weibull factors $k = 2.8$ and $C = 8.2$, with the given turbine data and using the WERA programme, we can find the capacity factor of the turbine at the site to be 0.335. Hence the annual energy production from the farm is 26411 MWh.

Life of the land based farm can be taken as 20 years. Hence, the Energy Payback Ratio is

$$EPR = \frac{26411 \times 20}{15859} = 33.3$$

Similarly, the Energy Payback Period of the farm is

$$EPP = \frac{15859}{26411} = 0.6 \text{ years}$$

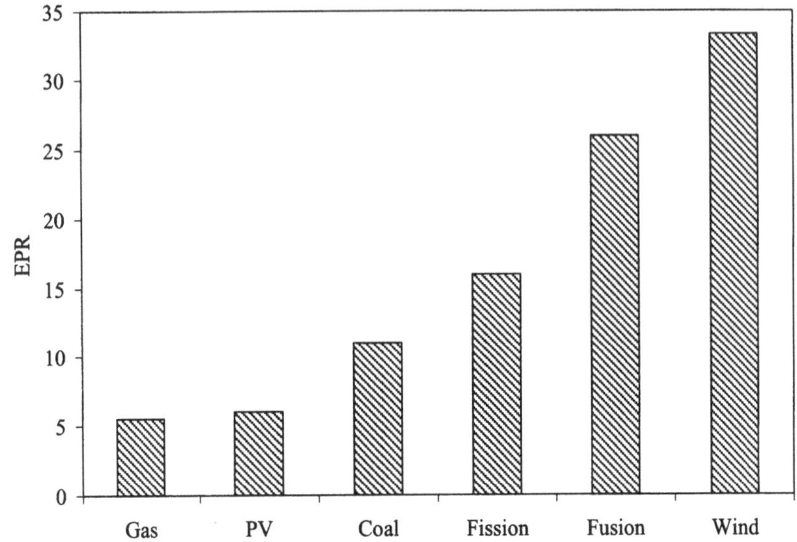

Fig. 6.4. Energy Payback Ratio of various energy sources

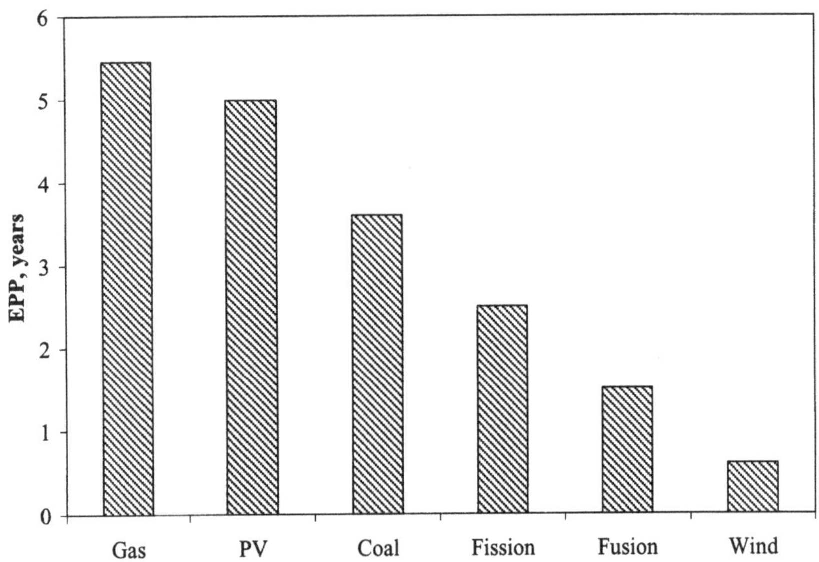

Fig. 6.5. Energy payback period of various energy sources

Hence, the wind farm would pay back all the energy consumed in its life cycle within 8 months of its commissioning.

The Energy Payback Ratio and Energy Payback Period of different sources are compared in Figs. 6.4 and 6.5. Here, the life period of the coal, fusion and fission projects are assumed to be 40 years and that for gas and PV technologies are taken as 30 years. For coal, gas, fusion and fission technologies, capacity factor of

75 per cent is assumed. Similarly, the conversion efficiency of PV modules are taken as 6 per cent. Comparing various technologies, we can see that wind energy has the highest EPR and lowest EPP, which is a clear indication of the environmental merit of wind energy over other options.

Example

Calculate the EPR and EPP of an offshore wind project with 9 turbines of 600 kW rated capacity. Take the same turbine specifications as given in the previous example.

With an energy consumption of 6173.6 GJ per system for offshore installation, the total energy input to the project during its life cycle is 55562.4 GJ, which is 15434 MWh. The life span of offshore systems is generally higher than the one on land. Let us consider a moderate life of 25 years. Assuming the same turbine and site characteristics, the project will generate 15846 MWh per annum. Hence,

$$EPR = \frac{15846 \times 25}{15434} = 25.66$$

Similarly,

$$EPP = \frac{15434}{15846} = 0.97 \text{ years}$$

The offshore project has lower EPR and higher EPP in comparison with the land based system. These differences are due to the extra materials used for the foundation of the offshore project.

6.2.2 Life cycle emission

Wind energy do not pose the threat of atmospheric emission during its energy generation phase. However, energy in various forms is being consumed during the construction, commissioning and decommissioning phases of the project. In the Life Cycle Emission (LSE) analysis, we account emissions during all these phases of the turbine's life cycle.

Possible emissions corresponding to the materials used for 600 kW turbine, assuming the Danish conditions, are given in Table 6.6, 6.7 and 6.8.

Example

Calculate the specific emissions during the lifecycle of a 600 kW land based wind turbine with specifications as in Example 2.

Consider the emissions of various forms corresponding to the 600 kW turbine given in Table 6.6. As seen from the above example, the turbine produces 1761 MWh/year from the site. Thus, the total energy produced by the turbine during its life of 20 years is 35220 MWh. Dividing the total lifecycle emissions by

Table 6.6 Emissions from different materials used for a 600 kW land based wind turbine

Materials	SO_2	NO_X	CO_2	N_2O	CH_4	NMVOC	CO
Steel	920.17	602.87	146370.49	4.44	2.54	10.15	59.02
Aluminum	37.38	23.14	6111.63	0.19	0.12	0.26	1.33
Copper	26.35	17.16	4836.64	0.14	0.12	0.19	1.16
Glass	1.15	3.18	766.92	0.01	0.05	0.20	0.87
Polyester and epoxy	54.98	35.30	9458.40	0.29	0.19	0.48	2.64
Reinforced iron	209.95	128.02	44841.60	1.30	0.86	2.59	22.61
Concrete	3.39	847.50	238317.00	0.00	0.00	0.00	0.00
Total	1253.38	1657.17	450702.68	6.37	3.88	13.87	87.62

Table 6.7 Emissions from additional materials required for the offshore wind turbine

Materials	SO_2	NO_X	CO_2	N_2O	CH_4	NMVOC	CO
Reinforced iron	209.95	128.02	44841.60	1.30	0.86	2.59	22.61
Concrete	3.39	847.50	238317.00	0.00	0.00	0.00	0.00
Copper	110.25	71.80	20235.46	0.59	0.50	0.77	4.86
Lead	73.34	79.91	11906.50	0.44	0.28	2.22	8.18
Steel	67.86	44.46	10794.42	0.33	0.19	0.75	4.35
Total	464.79	1171.69	326094.97	2.66	1.83	6.33	40.01

Table 6.8 Emissions due to the disposal of materials

Materials	SO_2	NO_X	CO_2	N_2O	CH_4	NMVOC	CO
Polyester and epoxy	8.64	14.40	11232.00	0.38	0.58	0.86	210.05
Miscellaneous	3.02	5.04	3931.20	0.13	0.20	0.30	73.52

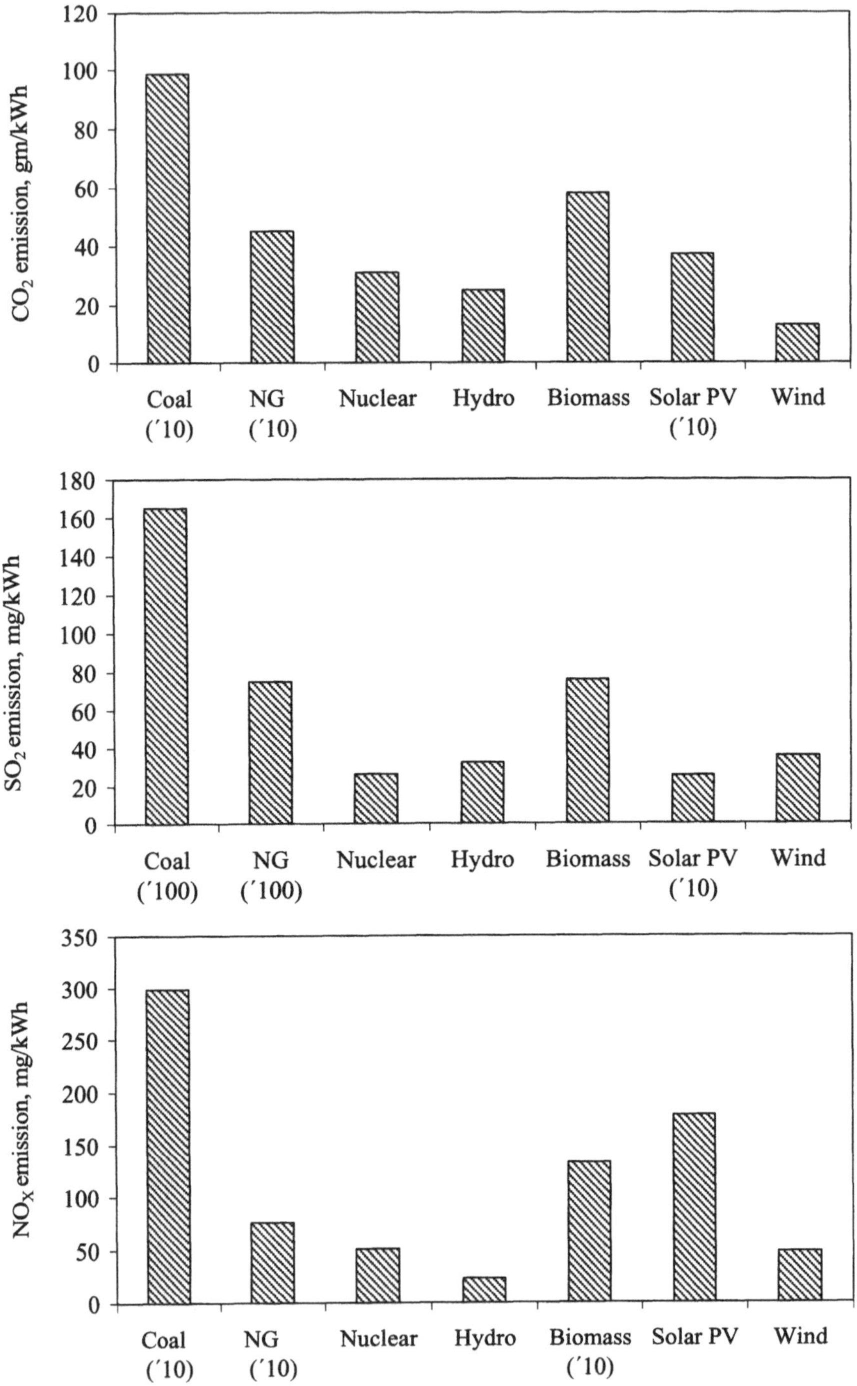

Fig. 6.6. Comparison of life cycle emissions from various energy sources

Table 6. 9. Specific emissions from a 600 kW wind turbine

Pollutant	gm/kWh
SO_2	0.035918223
NO_X	0.047603969
CO_2	13.22731062
N_2O	0.000195494
CH_4	0.00013228
NMVOC	0.000426834
CO	0.010539174

this lifecycle energy output, we get the emissions per kWh generated. The results are given below in Table 6.9.

We can see that the major pollutants here are CO_2, SO_2 and NO_x. Emission of N_2O, CH_4, CO and NMVOC are marginal. Life cycle emissions of wind energy, as compared with other sources are shown in Fig. 6.6. Even if we consider all possible emissions throughout its life cycle, wind emerges out as the cleanest source of energy available today.

6.3 Environmental problems of wind energy

Human intervention of any nature has its on environmental consequences. Wind energy is not an exception. Although wind is one of the cleanest sources of energy and does not pollute the environment with harmful gases during its energy conversion process, wind energy systems pose some environmental problems. The major environmental consequences of wind energy conversion are (1) risk to avian species (2) noise emission and (3) visual impacts.

6.3.1 Avian issues

Birds and bats may collide with wind turbines, as they do with any structures on their route. It is by the late 1980's that the impact of wind turbines on avian population has become an issue of concern. Studies conducted at the large wind resource area at the Altamont pass, located 90 km east of San Francisco, California, indicated high levels of bird mortality due to collision with wind turbines [11, 25, 26]. The species at stake were raptors and golden eagles, which are present round the year at this area.

Reports of this avian mortality from the Altamont pass has raised serious concerns to both the environmentalists and wind farm developers, in the ensuing years. Experiences from the Altamont pass was exaggerated and extrapolated. As a result, even the proposals for wind energy projects at other sites were guarded, fearing that the proposed future developments may also have similar impacts on the ecosystem. Several studies on the effect of wind turbines on avian population followed in the later years among which investigations conducted at the National Renewable Energy Laboratory are noteworthy [7, 29]. By now, it has been established that this is a site specific problem and the data related to a particular site, species, turbine and farm design cannot be extrapolated to characterize the wind turbine-bird interaction at other locations. Detailed studies have established that only a few potential wind sites around the world do have this risk to avian population.

In many cases, careful planning can avoid the risk to bird population due to wind turbines. Experiences from Altamont pass in US and Alcornocales national park in Spain [16] shows that sites which are well known habitat for rare bird species should be avoided, even if we have strong wind resources at such locations. Areas of intense avian activities can preliminary be identified with the help of local experts. Scanning available information on the area like literature and natural resource database can also be helpful.

Once the site for a wind plant is identified, the potential effect of the proposed wind farm development on the birds of concern should be systematically studied. If the available information in this regard is not sufficient to establish the biological suitability of the site, we should conduct onsite monitoring and surveys using appropriate methods and metrics. All the bird groups of special concern, including the breeding, migrating and wintering species, should be included in such studies. Apart from the mere number of birds and its distribution in the wind farm area, the functional behavior of the species in the region should also be included in the analysis.

The most common experimental design adopted in such investigations is the Before-After Control Impact (BACI) [10]. Under this method, data on avian activities are collected from the wind site before and after the construction of the farm as well as from a controlled reference site. The activities are recorded using systematically planned point count surveys. The bird fatalities are estimated through carcass searches around some systematically or randomly selected turbine areas. Although the true population size is not included in such studies, trends and indices that can be correlated with actual population parameters can be derived. National Wind Coordination Committee (NWCC) has prepared the procedures for conducting such investigations [2].

Results of such studies can be used to develop an index for risk of collision of birds with wind turbines. The collision risk index for a given bird species is given by [8]

$$I_R = U\ P_f\ P_t \tag{6.3}$$

Where U is the mean use by a species adjusted for visibility bias, P_f is the fraction of the total observations in which the activity of the subject is 'flying', and P_t is the fraction of all the flight height observations coming under the height band swept by the turbines.

For example, suppose the average observations of a given species recorded in a study is 0.20. Let the adjustment factor for visibility bias be 0.25, which means that a quarter of the birds actually present in the study is observed during the investigation. Thus the estimated use of the site by the birds, adjusted for visibility factor is 0.20/0.25 = 0.8. If the species is found to fly in the area for 70 per cent of time in which the 28 per cent of time the height of flight was within the height band swept by the turbines, then the risk index is given by

$$I_R = 0.8 \times 0.7 \times 0.28 = 0.16$$

It should be ensured that the risk index is under the limit permitted by the environmental laws framed by the regulatory bodies.

The avian risk at a given site can be reduced to some extent by the optimum layout of the wind farm and configuration of individual turbines. If the migration route of the birds and the corresponding climatic conditions are identified, the turbines can be laid out in such a way that clear spaces are available for the birds to pass through. In areas of migration, the turbines can be widely spaced leaving clear route for the migration and thus the risk of bird collision can be minimized.

Visual acuity of birds is a critical factor influencing the avian-wind turbine interaction. Considerable research in this aspect has been done by Dr. Hugh McIsaac [21]. He found that the raptors can see the turbine blades having an average width of 0.6 m, even from a distance of 1000 m. However, as the blades start rotating, the visibility reduces as a result of blurring. Contrast is another factor influencing the visibility. Inorder to make the moving components distinctly visible, the contrast between the blades and the background should be high. Thus to make the turbine blades maximally visible, the turbine blades may be painted in a highly contrasting pattern.

Noise can also be used as an effective tool for making the turbine more detectable for birds, especially during the night hours. Modern turbines are less noisy and the inherent noise level of these turbines does not keep the birds away. Efforts are even made to attach separate devices to the rotor tips which can generate noise in levels that can startle birds and draw their attention. Different types of bird flight diverters like balls and spirals as used in power lines are also being experimented.

Horizontal axis turbines mounted on horizontal lattice towers are more susceptible to bird collision. Further, these types of tower structures with flat crossbars provide ideal platform for perching birds. Tubular structures are less preferred by the birds for perching and hence could reduce the probability of wind turbine-bird interaction. Most of the modern wind turbines are provided with tubular structures.

Collision of birds with wind turbines is not as severe an issue as it is often being projected. It is reported that a bird in the vicinity of a farm will normally collide with the turbines for not more than once in every 8 to 15 years [6]. As a matter of fact, birds may collide with any man made structures. For example, around 1.25 million birds die every year due to collision with tall structures like buildings and towers. Similarly, 97.5 million birds are killed as they strike with glass. Estimates of bird mortality due to hitting vehicles are more alarming – 57 million a year [17]. In comparison with these causalities, wind turbines have only a marginal impact on avian population. Accessory structures and distribution networks often pose more threat to birds than the wind turbines itself. In a study spanned for over four years [13], it was found that fifty four per cent of the mortalities were due to the electric generation and transmission structures. Causalities due to the wind turbine blade striking were only 38 per cent.

While discussing the avian issues of wind turbines, it is logical to consider the effects of other generation technologies on the environment. For example, the green house emissions, acid rain and smog resulting from the coal based plants have more far reaching effects on the environment and the ecosystem, than the wind turbines. This already destroyed many lakes, streams and forests all around the globe. Changes in the weather pattern due to the global warming threaten many climate sensitive species. Apart from the risk from radioactive wastes, the hot water discharged from the nuclear plants has adverse effect on the marine habitat. For example, San Onofre nuclear station in California is responsible for killing around 21 tons of fish every year, including several billions of fish larvae [24]. A single accident from these plants may wipeout thousands and thousands of birds and other protected species from the earth. Structures of the conventional generating plants (towers, electric lines etc.) also pose threat to birds and bats, which are often left unaccounted.

Although relatively small numbers of birds are killed due to collision with wind turbines, the wind energy communities all around the world are making every effort to resolve this issue of avian mortality. For example, the American Wind Energy Association (AWEA), in coordination with the research organizations and industrial groups, has established an environmental task force to tackle this problem. Similarly, in UK, the Energy Technology Support Unit (ETSU) under the Department of Trade and Industry has initiated activities to reduce the avian deaths due to wind energy projects. Findings of several studies made in this direction is compiled in [19].

6.3.2 Noise emission

Like any other rotating mechanical systems, wind turbines also create some noise during their operation. The noise was a severe problem with the turbine designs in the 80's. During that time, environmental impacts of wind energy was not a matter of concern as it is today. Some of the turbines constructed during this period was quite noisy and could annoy people even at far away distance. In the later years, the noise emission from wind turbines has attracted more attention form the environmental groups, regulatory authorities and the wind turbine industry. A number of design modification followed in the preceding years and as a result, the modern wind turbine is a much quieter piece of machinery.

Any unwanted sound can be considered as noise. The noises generated by a wind turbine may be tonal, broad band, low frequency or impulsive. Tonal noises have discrete frequencies whereas the broad band noises have continuous frequency above a level of 100 Hz. The turbine may also produce noises with low frequency ranging from 20-100 Hz along with those due to momentary acoustic impulses. The magnitude of noise can be expressed either in terms of sound power level or sound pressure level. The sound power level indicates the acoustic power with which the noise is emitted from the source whereas the sound pressure tells us the intensity of noise experienced by the listener located at a given point.

The easiest way to identify the noise from wind turbines is to feel oneself 'how annoying' the noise is. For precise measurements, sound level meters are used. In theses devices, the acoustic variation causes changes in electrical signals which are then calibrated in terms of sound pressure levels for recording. Usually, all the frequencies are weighed and grouped into broad band level. Sound level meters are equipped with filters (A, B and C) in tune with the human response to change in noise. Filter A is most common in environmental assessments and the resulting measurement is expressed in dB (A). There are a few standards developed for the measurement of noise from wind turbines. The prominent standards are by the American Wind Energy Association (AWEA) [3] and the International Electro Chemical Commission (IECC) [16]. The latter is widely used in many parts of Europe and America. Under this standard, the sound power level is measured at 10 m height at a referral wind speed of 8 m/s.

Two types of noises are generated from a wind turbine-mechanical noise and aerodynamic noise. The mechanical noise is contributed by the relative motion of components like gear box, generator, yaw motors, cooling fans, hydraulic pumps and other accessories. For example, noise emission from the gear box of 2 MW wind turbine is in the tune of 97.2 dB (A) where as the contribution from the generator is 87.2 dB (A). The auxiliary components may emit noise at a level of 76.2 dB (A) [27]. Changes in the design of these components can significantly affect the frequency and tone of the noise. Mechanical noises are mostly of tonal type.

Flow of air around the blade causes the aerodynamic noise. When the blades of the wind turbine interacts with the air stream, a number of complex flow phenomena occurs around the blade, each contributing to the noise generated from the system. The aerodynamic noise is mostly of broad band type with some amount of low frequency or even tonal components. The typical aerodynamic noise from a 2 MW turbine can be at the level of 99.2 dB (A) [27]. Shape of the blade tip, tip speed ratio, pitch setting, trailing edge thickness, blade surface finish and twist distribution are some of the factors affecting the aerodynamic noise. A more rigorous analysis of the aerodynamic noise, critically analyzing various contributing factors can be found in [33].

Frequency and magnitude are the two major factors defining the characteristics of sound. In general, human beings can hear the sound in the frequency range of 20 Hz to 20 kHz. The threshold level of human tolerance towards noise is 140 dB (A) [27]. However, apart from the acoustic characteristics, attitude also plays an important role in human tolerance to any noise. The same sound which may be delightful to one person may be a disturbing noise for the other. The negative attitude towards a source may make the noise from it more aggravating, even though the intensity is low.

The intensity of the noise from a turbine increases with the wind speed. However, as with the increase in wind speed, the background noise due to the wind (from trees, bushes etc.) also increases, which in turn pacifies the effect of noise generated from the turbine. Thus, to a certain extend, the noise produced by wind turbines are masked by the wind itself (interestingly, this wind induced background noise often make the precise measurement of wind turbine noise difficult. This 'pseudo' noise problem is overcome by either providing a wind shield to the measuring microphones or making the measurements with two microphones and eliminating the effect of wind using correlation techniques). Other noises (traffic, industrial or agricultural) may also mask the noise created by the turbine.

Down wind turbines are likely to be noisier than the upwind design. The rotors of these turbines operate downwind to the tower. In this case, the wake vortices generated by the tower and the aerodynamic lift of the rotor blades interacts, resulting in impulsive noise of low frequency. Similarly, the pitch controlled turbines generate lesser noise than the other designs. The speed of the blade tip is a critical factor in deciding the aerodynamic noise generated by the system. To minimize the noise level, it is advisable to restrict the tip speed well within the limit of 65 m/s [18]. Smaller wind turbines tend to have higher rotational speeds. Many times, it may not be practical to restrict the tip speed of these machines within this limit. Power regulation in these systems (by deviating the rotor from high winds) may also cause noise problem. Hence, smaller wind turbines may felt to be noisier for their size.

The intensity of noise from the turbine, felt at a point of interest (residential areas, offices etc) primarily depends on the distance between the source and the point. As the distance between the source and the receiver increases, the pressure level decreases. Propagation path is another factor deciding the noise intensity. For a flat terrain without obstructions, it is logical to assume a hemispherical path for the noise propagation. Further, the ground characteristics like vegetative cover, surface roughness and climatic factors like temperature, humidity, wind duration etc. may also affect the noise propagation.

Elaborate methods, considering all these effects, are available for modeling the noise emission and propagation from wind turbines [14]. Commercial wind energy softwares like the WindFarm© have provisions for noise propagation analysis.

For a relatively simpler estimate, we can consider the noise propagation path to be hemispherical. With this assumption, the sound pressure level L_P at a distance R from a wind turbine, radiating noise at an intensity of L_W is given by

$$L_P = L_W - 10 \log_{10}\left(2\pi R^2\right) - \alpha R \tag{6.4}$$

where α is the sound absorption coefficient. In case of broad band sound, α may be taken as 0.005 dB (A)/m.

Example

The noise generated by a wind turbine with 50 m tower, measured at the source is 104 dB (A). Calculate the level of noise due the turbine at a residential area located 200 m away from the turbine. Assume a sound absorption coefficient of 0.005 dB (A)/m and neglect the effect of ambient noise.

From the above expression, the sound pressure level at 200 m is

$$L_P = 104 - 10 \log_{10}\left(2\times\pi\times 200^2\right) - 0.005\times 200 = 49 \text{ dB(A)}$$

The levels of noise from the turbine considered in the above example, at different distances are shown in Fig. 6.7. We can see that the noise emitted from the turbines should not be a matter of great concern, if the residential areas are not very close to the wind farm. For example, the noise level at 100 m is around 55 dB (A)

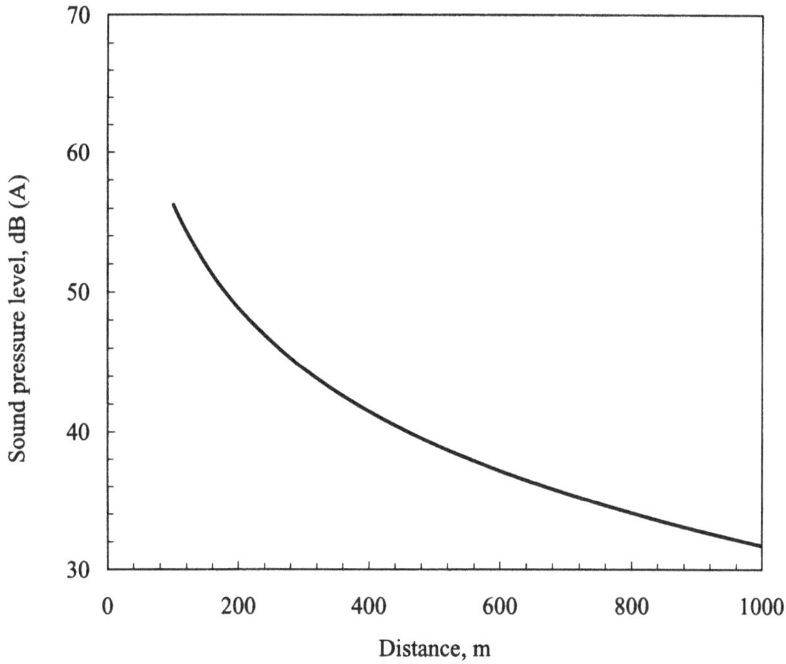

Fig. 6.7. Noise from a wind turbine at different distances

in this case, which is equivalent to the noise from a car passing at 40 mph at this distance [4]. Residential areas are usually located at least say 350 m away from the farm. At this distance, the intensity of noise from the turbine is only 43 dB (A).

In practical situations, we may be interested to know the effect of noise at a point due to two or more different sources. For example, if we want to calculate the acoustic effect at a residential area from a wind farm having a number of turbines, the combined effect of noise generated by all these turbines at the point of concern is to be calculated. The noise level due to the individual turbines at this point can be calculated using the above expression. However, as the noise levels cannot be added linearly, we have to compute the sound power level of different sources as experienced at this point to calculate the resulting noise. For a given noise level of L_P, the sound power P_N can be approximated as

$$P_N = 10^{\left(\frac{L_P - 90}{10}\right)} \tag{6.5}$$

Thus, if the noise level at a point is 50 dB(A), then the corresponding power level is 0.0001 W/m^2.

To estimate the combined effect of noise from two or more sources at a point, we have to add the individual power corresponding to these noises and then find out the noise level corresponding to the total sound power. From Eq. (6.5), taking logarithm and rearranging , we get

$$L_P = 10 \log_{10}(P_N) + 90 \tag{6.6}$$

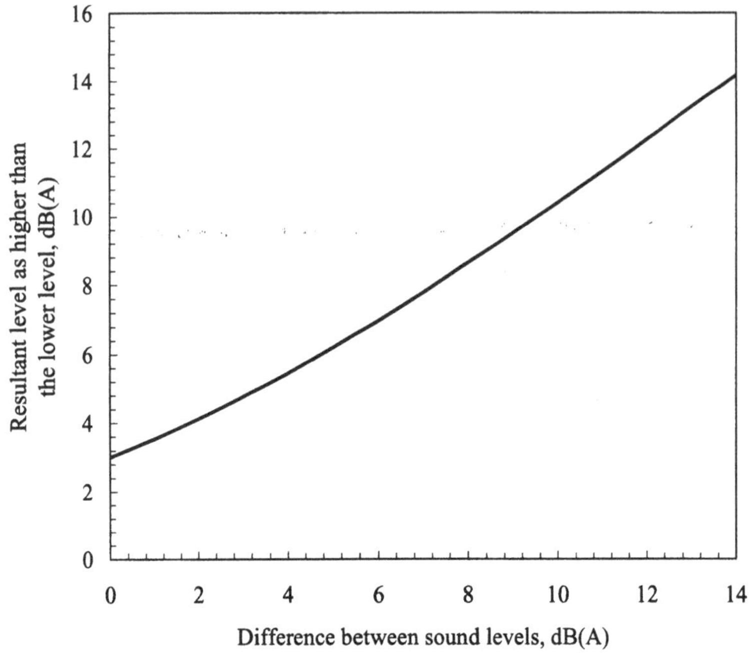

Fig. 6.8. Combined effect of noise from two sources at a point

Example

Two wind turbines emitting 104 dB (A) at the source are located at 200 m and 240 m away from a residential point. Calculate the combined acoustic effect of these turbines at the residence.
As seen from the above example, the noise level at the residence due to the first turbine is 49 dB (A). The corresponding power level is

$$P_N = 10^{\left(\frac{49-90}{10}\right)} = 7.94\times10^{-5}\ \text{W/m}^2$$

Similarly, for the second turbine, the noise level is

$$L_P = 104 - 10\ \log_{10}\left(2\times\pi\times240^2\right) - 0.005\times240 = 47\ \text{dB(A)}$$

And the corresponding power level is

$$P_N = 10^{\left(\frac{47-90}{10}\right)} = 5.01\times10^{-5}\ \text{W/m}^2$$

So, the total sound power is 0.00013 W/m^2. The noise level, corresponding to this power level is given by

$$L_P = 10\times\log_{10}(0.00013) + 90 = 51.12\ \text{W/m}^2$$

Effect of noise from two different sources at a point is shown in Fig. 6.8.

Difference between the noise levels, as individually experienced at the point of interest, is indicated in the X axis. Y axis shows the difference between the resultant noise and the lower one.

We can see that while two noises of the same intensity are added (difference 0), the resultant noise will be 3 dB(A) higher than the lower noise. Similarly, as the difference between the noises to be added are high (for example 12 dB(A)), then the lower noise will not have any significant effect as the resulting noise level will be almost equal to the noise of higher intensity.

Growing concern over the noise emission has made the quite operation of the turbines an important criterion for the successful design of wind energy conversion systems. Turbine designs are being continuously improved to reduce both the mechanical as well as aerodynamic noises. The mechanical noises are mainly reduced by modifications in gears and transmission components. Especially finished gears with slightly flexible semi-soft core and hard surface are being used with turbines to reduce noise. The cooling fans used in some of the modern designs operate at low speeds to reduce the mechanical noise. Moving parts like generators, gears etc., mounted on the nacelle, are acoustically isolated from the tower and blades to reduce the noise propagation.

The major step in reducing the aerodynamic noise is to lower the operating tip speed ratio. However, running the system at lower tip speed ratio may affect the power curve, resulting in reduced energy generation. Opting upwind turbines over the downwind design, lowering the angle of attack of the blades and special blades with tailing edge are some other means to reduce the noise. With improved blade profile, the aerodynamic efficiency of blades is being enhanced and thus more energy is utilized for energy generation than wasted in the form of noise. Providing aerodynamic shape to projected parts of the structure like nacelle and towers has also proved to be effective in reducing the aerodynamic noise.

Systematic acoustic analysis during the project planning stage can avoid possible noise annoyance from wind turbines. If the noise generated from a particular design of turbine, proposed to be installed at a site, is available form the manufactures, it is possible to model the propagation of noise on the basis of climatic and terrain factors. The pre-construction surveys can tell us about the background noise prevailing at the proposed site and thus we can determine how much noise is going to get added to this due to the presence of the wind turbines. This helps in ensuring that the noise emitted from the turbines is safely within the regulatory limits.

How much noise can be permitted from a wind turbine? Answering this question is not simple. The permitted noise level may depend on the background noise already existing at the wind farm site. According to the World Health Organization (WHO) norms, the allowable noise level at the out doors of a residential area is 40 dB (A)[30]. Most of the countries have strict regulations on the levels of noise emissions permissible in residential localities. The limit however varies from country to country [31]. For example, the highest permissible noise level from wind turbines in Sweden is 40 dB(A) where as the Oregon Department of Energy (USA) may permit a night time noise upto the level of 50 dB(A). In Denmark, where special legislation is framed for wind turbines, the limit for outside dwelling is 45 dB(A) where as for sensitive areas it is restricted to 40 dB(A).

German regulations restrict the noise level in residential areas to 50 dB(A) during the day and 40 dB(A) at night. Some countries like the Netherlands follow a more scientific approach in noise regulation. Here, the permitted noise at night vary with the wind speed, starting from 40 dB(A) at 1 m/s, upto 50 dB(A) at 12 m/s. The corresponding day and evening time limits starts at 50 dB(A) and 45 dB(A) respectively.

Noise pollution from wind turbines is not a serious issue as it is being considered. In most of the cases, the noise from wind turbines does not exceed the limits put forth by the permitting agencies. For example, a wind turbine with moderate sound power emission may be as quite as a kitchen refrigerator at a residential area 300 m away from the source. Apart from annoyance in few cases, no health problems are reported due to the wind turbine noise. A study conducted in Europe, incorporating sixteen sites from Denmark, Germany and Netherlands showed that only 6.4 per cent of the residents felt that the noise from wind turbines as annoying [34]. With proper planning and improvements in the design, the noise emission from wind turbines will further be reduced in the coming years.

6.3.3 Visual impact

Another environmental concern of wind farm development is its impact on scenic beauty of the landscapes. Wind turbines are tall structures installed in open areas which make them visually prominent in the landscape. The turbines may dominate our sight upto 2 km or even more. They are often felt to be an important element in the landscape evenup to a distance of 5 km. As "beauty lies in the viewer's eyes", some may like the sight of a wind turbines, generating energy in an environment friendly way near to their area. However, we must not forget that there may be some people who might consider the turbine as 'a box on a long stick' erected to ruin the scenic beauty of the landscape. Hence, the turbines should be naturally integrated to the landscape to make them visually attractive and acceptable.

The value that one assign to the landscape and its surroundings is an important factor in molding his opinion on the wind farm. The aesthetics value of a landscape is judged in terms of its visual, historical, ecological, socio-cultural, religious and mythological importance. Hence it is advisable to assess the sensitivity of the landscape towards these factors, before going for a wind farm project. There are several methods to establish the aesthetic sensitivity of a landscape [22]. Use of these techniques in conjunction with the public opinion surveys can give us an indication on the appropriateness of a site for wind farm installation. Many local permitting authorities have already quantified the sensitivity of their landscapes using these methods, which are available for the developers for assessing the suitability of their project.

Apart from the characteristics of the location where the wind farm is proposed, the visibility of the location from other key points is also important. For example, peak of the mountains may be visible from a far away distance, whereas the ridges may be less prominent. Hence an undifferentiated and horizontal ridgeline can be a better site for placing the wind turbines. Similarly landscapes which are highly contrasted with open meadows, long raised strips, rock cliffs, ponds etc. may have high scenic value. These distinct and varied profiles may also be visible from long distances. An object draws more attention, even from a distance, if it is placed in the focal point of a scenic view. Hence, it is preferable to place the wind turbines on the periphery of scenic spot so that they would not compete with the surroundings for viewer's attention.

It has been established that small clusters of turbines are more acceptable than long arrays. A large project comprising of a number of turbines fits comfortably into the landscape if they are sub grouped into several visually comprehensive units and arranged in a well defined geometric pattern-preferably linear. Farms with turbines scattered all over the landscape may lack in coherency and clarity. The number of turbines included in a cluster may vary from case to case depending on the landscape features but normally 6-10 turbines can make a compact unit. It is preferable to have uniform spacing between the turbines in 'farm land' landscapes to have a consistent image. However, in mountain moorlands, multiple arrays of randomly spaced turbines may not cause any visual annoyance. Clusters look better if the turbines are tightly packed in it (with sufficient spacing to avoid the interference between the turbines).

Turbines of same type, size and equal number of blades are preferred for clean, simple and repetitive view. Uniform tower height is also advisable. All the turbines should be functional with rotors rotating preferably in the same direction and at the same speed. Mismatch between the turbines in a cluster may result in chaotic effect. Tubular towers are preferred over the lattice towers. The turbine components should be aesthetically engineered with all its major components - tower, nacelle and blades - proportionally sized. Simple and aerodynamically shaped components can be aesthetically more appealing. All the turbines in a farm should be uniformly colored. It is preferable to give same color to the blades, nacelle and towers as it avoids color contrast between the components. White color is preferred, although some designers go for off-white and grey options.

Structures built up for substations, transmission lines and access roads often cause more visual annoyance than the wind turbines itself. As far as possible these structures may be buried so that they may not come in direct view. Growing vegetations, which are matching with the natural settings of the surroundings, is another possible way to hide substations. Similarly, the transformer boxes can be hidden at the tower base. Additional transmission lines needed for the farm can be minimized by locating the generating facilities near to the existing power lines. The roads and other structures made for the construction of wind farm should be temporary in nature and the cleared area may be re-vegetated after the construction.

Fencing and barbed wires around the wind farm may be avoided to make the turbines and surrounding areas accessible to the public. This is helpful in eliminating the public anxiety about the project and clarifying any possible doubts. Further, interacting with the turbines will give them a chance to feel the way in which clean energy is being generated at their area. It is a common experience that, once the project is established and the environmental benefits of wind energy are understood, wind farms become visually more acceptable.

Fig. 6.9. Computer simulation of visual impact of wind turbines on landscape using the WindFarm Software© (Courtesy of ReSoft, UK, www.resoft.co.uk)

Carefully designed and cleverly constructed wind farms can be a positive addition to the landscapes. Wind farms with layouts, naturally blending with the prevailing scenery and turbines designed with aesthetic sense can add to the scenic beauty of the landscape.

Computer models are available to assess the visual impact of wind farms on the landscape and the flora and fauna, before the farms are actually constructed. The WindFarm© software has a special photomontage module using which the pictures of the proposed landscape and the turbines to be installed can be superimposed to simulate the visual impact of the farm. View of a landscape before and after the construction of a wind farm simulated using the software is shown in Fig. 6.10. The software can also animate the turbines to conceptualize the view while the turbines are actually generating. More exhaustive discussions on the integration of wind turbines with landscape are available in the recent reports by the Danish Energy Agency and Heriot-Watt University [23,32].

During its rotation, blades of the turbine may cast shadows intermittently, which results in flickering effect on the surrounding areas. This 'disco' effect may cause annoyance to the people residing very close to the farm. The intensity of shadow casting depends on the rotor speed and direction, number of bright sunshine hours and the geographical location of the installation. Estimating the shape, location and time of the shadow, based on all these factors is a tedious process for which computer programmes are being used. Shadow casting is not a very serious problem in wind farm management as this will affect only a small area, very close to the turbine. Interference with TV and radio signals was a problem with earlier turbines with metallic blades. However, with the introduction of composite blade materials such as fiber glass and plastics, this problem is minimized in modern turbines.

References

1. Allam RJ, Spilsbury CG (1992) A study of the extraction of CO_2 from the flue gas of a 500 MW pulverized coal fired boiler. Proceedings of the 1st International Conference on Carbon Removal (FICCDR), Energy Conversion and Management, Pergoman Press
2. Anderson R, Morrison M, Sinclair K, Strickland D (1999) Studying wind energy/bird interactions: A guidance document. Metrics and methods for determining or monitoring potential impacts of birds at existing and proposed wind energy sites. National Wind coordination committee (NWCC), Washington DC
3. AWEA (1989) Procedure for measurement of acoustic emissions from wind turbine generator systems. TeirI – standard 2.1, American Wind Energy Association
4. AWEA (2000) Facts about wind energy and noise. American Wind Energy Association, Washington D.C.
5. AWEA (2002) The most frequently asked questions about wind energy. American Wind Energy Association, Washington D.C
6. Benner JBH (1992) Impact of wind turbines on bird life: An overview of existing data and lacks in knowledge inorder of the European community. Final report, Consultants on Energy and Environment (CEE), Rotterdam, The Netherlands
7. Dooling R (2002) Avian hearing and avoidance of wind turbines. NREL/TP-500-30844, National Renewable Energy Laboratory, Colorado
8. Erickson WP, Strickland MD, Johnson GD, Kern JW (2000) Examples to statistical methods to assess risks of impacts to birds from wind plants. Proceedings of National Avian-Wind Power Planning meeting III, San Diego, California pp 172-182
9. Fenhann J, Kilde NA, Runge E, Winther M, Boll Illerup J (1977) Inventory of emissions from Danish sources, 1972-1995. Riso National Environmental Research Institute

10. Green RH (1979) Sampling designs and statistical methods for environmental biologists. Wiley, New York, NY
11. Howell JA (1997) Avian mortality at rotor swept area equivalents, Altamont Pass and Montezuma Hills, California. Trans. Western Sector Wildlife Society, 33:24-29
12. http://www.greenontario.org
13. Hunt WG (2000) A population study of Golden Eagles in Altamont Pass wind resource area: Population trend analysis 1994-1997-Executive summary. Proceedings of National Avian-Wind Power Planning meeting III, San Diego, California pp 15-17
14. IEA (1994) Expert group study on recommended practices for wind turbine testing and evaluation, 4. acoustic noise emission from wind turbines. International Energy Agency
15. IEA (2003) Emissions. Key world energy statistics. International Energy Agency pp 44-46
16. IEC (2001) Wind turbine generator systems-Part 11 Acoustic noise measurement techniques. IEC 61400-11, No. 88/141/CDV, International Electrochemical commission, Geneva, Switzerland
17. Kenetech Wind power (1994) Kenetech Wind power avian research programme update. Kenetech Wind power, Washington D.C.
18. Klung H (2002) Noise from wind turbines-standards and noise reduction procedures. Forum Acusticum 2002, Sevilla, Spain
19. Lowther SM, Tyler S (1996) A review of impacts of wind turbines on birds in the UK. Report No. W/13/00426/REP3, Energy Technology Support Unit (ETSU)
20. Marti Montes R, Barrios Jaque L (1995) Effects on wind turbine power plants on the avifauna in the Campo de Gibraltar region. Spanish Ornithological Society.
21. Morrison ML (2000) The role of visual activity in bird-wind turbine interactions. Proceedings of National Avian-Wind Power Planning meeting III, San Diego, California pp 28-30
22. MosArt Associates (2000) Landscape assessment for wind farm planning and design-character and sensitivity. Final report, Altener project AL/98/542 pp 1-18
23. Neilson FB (1996) Wind turbine and the landscape: Architecture and aesthetics. Birk Nielsens Tegnestue, Soendergade, Denmark
24. NIRS (1989) Committee finds massive sea life kills from San Onofre. Nuclear Information and Resource Service Groundswell 11 (2&3)
25. Orloff S, Flannery A (1992) A continued examination of avian mortality in the Altamont Pass wind resource area. Report P700-96-004CN, Ibis Environmental Services and BioSystems Analysis Inc.
26. Orloff S, Flannery A (1992) Wind turbine effect on avian activity, habitat use, and mortality in Altamont Pass and Solano Country wind resource areas. Report P700-92-001, BioSystems Analysis Inc.

27. Rogers AL, Manwell JF (2002) Wind turbine noise issues – A white paper. Renewable Energy Research Laboratory, Centre for Energy Efficiency and Renewable Energy, Department of Mechanical and Industrial Engineering, University of Massachusetts at Amherst, Amherst
28. Schleisner L (2000) Life cycle assessment of a wind farm and related externalities. Renewable Energy 20: 279-288
29. Schmidt E, Piaggio AJ, Bock CE, Armstrong DM (2003) National wind technology centre site environmental assessment – Final report. NREL/SR-500-32981, National Renewable Energy Laboratory, Colorado
30. Schwela D (1998) WHO guidelines on community noise. Proceedings of ICBEN 98, Sydney, pp 475-480
31. SEPA (2003) Noise annoyance from wind turbines-A review. Swedish Environmental Protection Agency, Report 5308
32. Stanton C (1996) The landscape impact and visual design of wind farms. ISBN 1-901278-00-X, Heriot-Watt University, United Kingdom
33. Wnager S, Bareib R, Guidati G (1996) Wind turbine noise. Springer Verlag, Berlin
34. Wolsink M, Sprengers M, Keuper A, Pedersen TH, Westra CA (1993) Annoyance from wind turbine noise in sixteen sites in three countries. European community wind energy conference, Lubeck, Travemunde pp 273-276

7 Economics of wind energy

In the previous chapters, we have discussed the technological aspects of wind energy conversion. The criteria for selecting the optimum system for a given location have also been described. However, in practice, selection of an energy option for an application is not made merely on the basis of its technical feasibility. The economic aspect of energy generation also plays a key role in the decision-making. Thus, along with issues like "how efficient is the system?" and "how much energy will it produce?", the question "at what cost can we generate energy?" is also relevant in choosing a source from the available options. This means that the project is to be optimized for the lowest possible cost per kWh generation.

One of the major factors limiting the wide spread acceptance of many renewable energy technologies is the high generation cost. However, with today's technology and institutional support, wind energy is economically competitive with other conventional sources like coal and natural gas. It is cheaper than all other renewables like solar, hydro, biomass and geothermal. With rigorous research and developmental efforts, cost of wind-generated electricity is further falling down. For example, during the past two decades, the cost of wind energy has dropped by more than 80 per cent. It is expected that this trend will continue in the future years also [2].

Economic issues of wind energy systems are multidimensional. There are several factors that affect the unit cost of electricity produced by a wind turbine. These may vary from country to country and region to region. Economic merit of a wind powered generation plant heavily depends on the local conditions.

For a wind turbine, the fuel is free, but the capital investment is high. While assessing the initial investment for the project, apart from the cost of the wind turbine, investment for other essential requirements like land, transmission lines, power conditioning systems etc. should also be accounted.

For making an intelligent investment decision, the net financial return from the project that is being planned for a given energy profile is to be assessed. For this, we have to estimate the costs involved in the generation as well as the benefits expected from the project. Computing the cost is relatively simple and straightforward as it can be determined by adding the fixed and variable costs. However, assessing the benefits is rather a complex process, as the value of electricity generated is influenced by several factors related to the local energy industry.

In this chapter, let us examine the major factors influencing the economics of wind energy projects. Methods to estimate unit cost of wind-generated electricity will also be discussed.

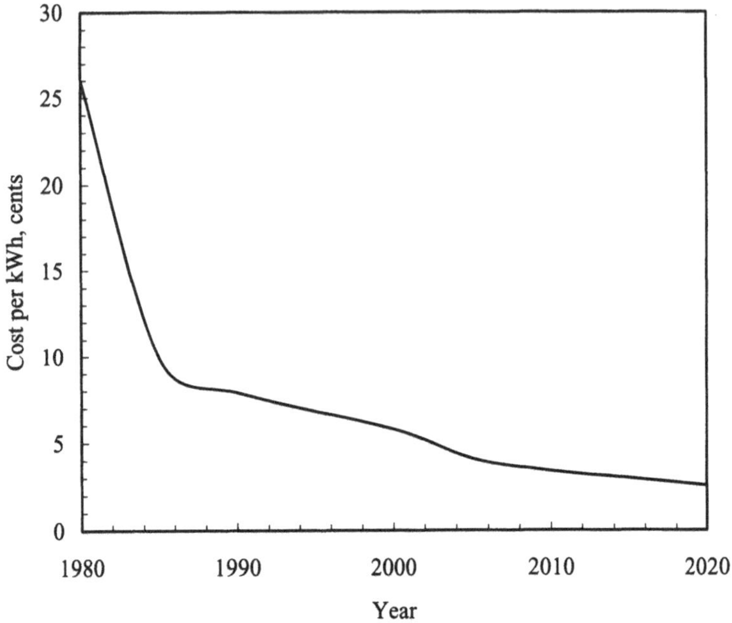

Fig. 7.1. Historic cost of wind energy

Further, based on the principles of engineering economics, criterion for assessing the financial merit of such projects would be evolved. We will adopt the 'present worth' approach in our discussions.

7.1 Factors influencing the wind energy economics

Cost of wind-generated electricity over the years from 1980 to 2020 is shown in Fig. 7.1. The estimates from 2005 to 2020 are projected, following the prevailing trend. We can see that the cost/kWh has fallen from 26 cents in 1980 to nearly 5 cents in 2005 - a reduction of 84 per cent in the last 23 years. The decline in the cost is more prominent during 1980 to 1985, which is attributed mainly to the increase in turbine size. Projections for the future indicate that the cost would further be reduced to 2.6 cents/kWh by 2020.

There are many factors that affect the economic viability of a wind energy project. They can broadly be grouped as site-specific factors, machine or system parameters, market factors and policy issues. Let us discuss these factors in brief.

7.1.1 Site specific factors

Energy available in wind spectra is proportional to the cube of the wind speed. This implies that when the speed of the wind at a location doubles, the energy increases by eight times. Hence, the strength of the wind spectra

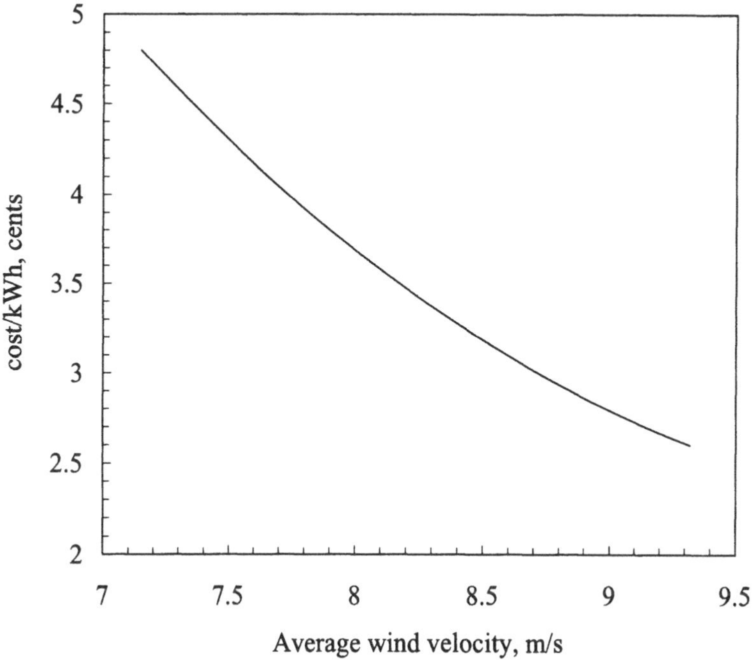

Fig. 7.2. Effect of wind velocity on the cost of wind generated electricity

available at the project site is one of the critical factors deciding the cost of wind-generated electricity. A typical example, demonstrating the effect of the average wind velocity at a site on the cost of unit energy produced, is shown in Fig. 7.2. When the average velocity increases from 7 m/s to 9.5 m/s, the cost is reduced by 50 per cent (from 5 cents to 2.5 cents per kWh). As illustrated in Chapter 5, matching between the wind profile at the site and the machine requirements is also important in keeping the generation cost to a minimum level.

Cost of land, installation charges and labor wages vary from place to place. Expenditure on foundation depends on the strength of soil profile as well as the extreme loads expected at the site. In case of grid connected systems, a major concern would be the distance from turbine to existing grid as the cost of developing additional transmission network should also be taken into our calculations. Extending access from wind farm to existing highways would also contribute to the cost, which may vary from site to site.

As the wind velocity increases with height, systems with taller towers generally produce more power. Towers are one of the costly items in a wind energy system. The minimum tower height required is determined by the surface roughness of the local area, which is again a site-specific factor. Local climatic conditions also influence the wind energy economics. High turbulence of wind at the site may demand more attention to the rotor. Further, presence of corrosive and other harmful substances in the atmosphere reduces life span of the turbine. Frequent maintenance may be necessary due to these factors, which in turn would increase the system's operational and maintenance costs.

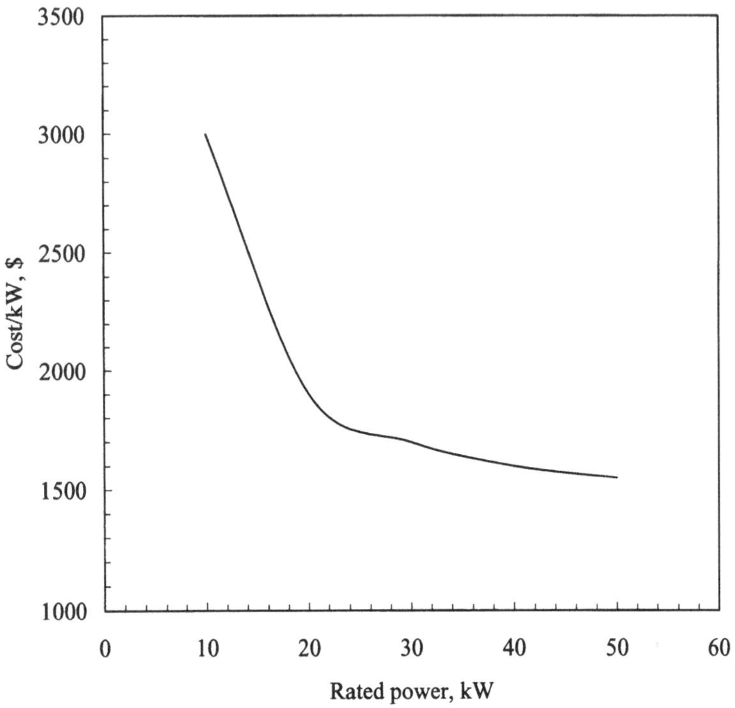

Fig. 7. 3. Cost reduction through scaling up

7.1.2 Machine parameters

Cost of the wind turbines can be considerably reduced by scaling up the system size. This means that the cost per kW of a 2 MW machine is lower than that of a 2 kW unit. Unit cost of wind turbines dropped from $ 2500/kW to $ 750/kW in the last 20 years. This cost reduction is achieved mainly through scaling up the turbine size. Thus, transition of wind energy technology from small units in the earlier days to the MW capacity machines today, has resulted in reducing the cost of wind-generated electricity.

In Fig. 7.3, unit cost of small-stand alone wind turbines of different sizes are compared. By scaling up the size from 20 kW to 50 kW, the cost/kW is reduced by 18 per cent (costs given are the average cost prevailing in the market, as different manufactures charge different rates for systems of same size). The reason is, expenditure on many components does not scale up at the same rate as the turbine size. For example, cost of safety features, electronics circuits etc. required for smaller units and larger machine are almost equal. Similarly, the manpower required for developing the turbine also does not vary much with the size.

Scale of production is another factor affecting the unit cost of wind turbines. The cost of production of the first few units will naturally be higher as the investment on research and development, as well as other infrastructure facilities is to be shared by this limited number of turbines.

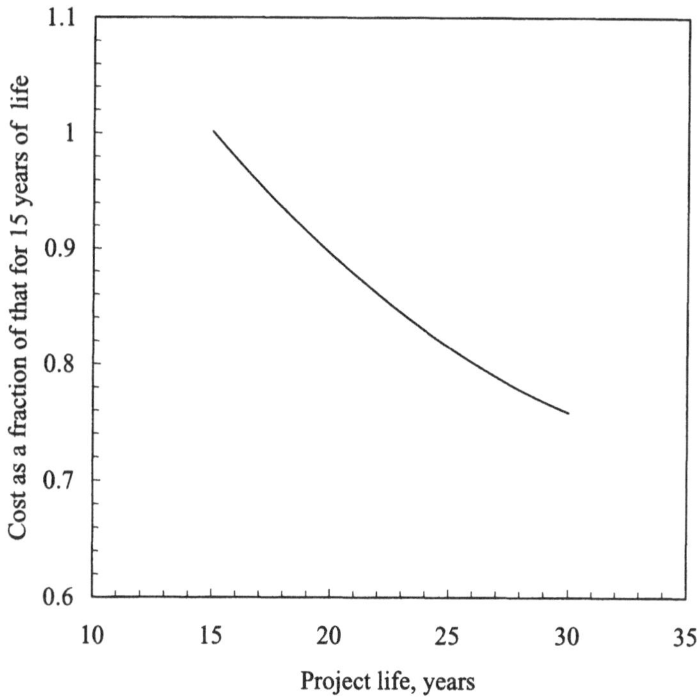

Fig.7.4. Effect of project life on wind generator economy

As the production rate goes up, unit cost comes down. The so-called learning curves, which are extensively used in the industry, can be used to correlate the production rate and unit cost of wind turbines.

Commercial manufactures reduce the cost of their turbines through typical design improvements. An example is the variable-speed constant-frequency machine. In these systems, cost of transmission components is reduced by smoothing up the load to be transmitted. At the same time, the aerodynamic efficiency of the rotor is fully exploited due to variable speed option. This will further make the grid integration simpler and efficient.

Economic life span of the turbine influences the cost calculations of wind energy systems. Generally, the life of a wind turbine may vary from 20 to 30 years [3,8]. When we design our system to sustain longer, the initial investment of the project will be distributed over more number of years, which in turn would reduce the annual cost of operation. A typical example is shown in Fig. 7.4. When the project life increases from 15 years to 30 years, the cost falls almost by 25 per cent. However, it must be ensured that the life period taken for our calculations is realistic.

7.1.3 Energy market

Existing energy market often decides the benefits of wind energy. If the electricity generated from the turbine is completely consumed by the owner, the economic advantage of installing the turbine is decided by the local electricity tariff. However, if the energy produced is in excess to his demand, surplus electricity may be sold to local utility company. Quite often, the rate at which the utility company buys electricity from individual turbine owners is much lower than their local retail-selling price. This is because the company's operational and management expenses as well as the profit are included in the retail tariff.

The rate at which utilities pay for the electricity may depend upon the time of the day at which the power is available. The companies have their peak energy demand in some particular hours of a day. If the wind turbine can supplement them in meeting this peak demand, the companies may pay higher price for the wind-generated electricity. The huge investments involved in enhancing their installed capacity in tune with the peak demand may thus be avoided and thus, they can afford to pay more for wind energy.

There are several special arrangements made for utilizing excess electricity generated by individual wind turbines. For example, under the system of net billing, the turbine owner can feed electricity to the grid when it is surplus and extract it back, when his demand is more. If the generation exceeds the consumption, the utility company will pay him for the difference at a pre-negotiated rate. The entire electricity inflow and outflow are metered.

7.1.4 Incentives and exemptions

With a view to promote clean and locally available energy sources, several federal and state governments are extending financial support to renewables in terms of exceptions and incentives. These may be in direct or indirect forms. Such supports make the wind energy option more attractive.

Financial supports for the wind energy projects are justifiable. It may appear that the energy market is open and competitive, that is a level playing ground for all the technologies. But it is not truly so. Many conventional energy sources like coal and natural gas enjoy subsidies in hidden forms. For example, in US, the Government have already spent more than $ 35 billion over the past 30 years for the treatment of coal miners who suffer from black lung disease [2]. This is definitely a hidden financial support to coal.

The fossil fuel based power plants load the atmosphere with pollutants. The cost of scavenging the atmosphere to get rid of these harmful chemicals should also be added to the cost of energy produced through these technologies. If all these hidden financial favors are accounted, the cost of many conventional sources will appear to be higher than they are claimed to be today. For example, the electricity generated from coal-fired plants would be 50 to 100 per cent more expensive than it is today, if environmental costs are included [7].

Energy policies existing in many countries are still in favor of the traditional technologies. Hence, a promising technology like wind may require certain degree of favor in its initial stages of growth. Keeping this in view, incentives of different forms are being extended to wind energy projects in many parts of the world.

These incentives can broadly be classified as investment incentives, production incentives and policy incentives.

An example for the investment incentive is the accelerated depreciation admissible in several countries. Usually, investors are permitted to deduct the annual depreciation of their expensive equipments and systems, from their income tax. This depreciation is calculated at a fixed rate, which is decided on the market value of the equipment. In case of wind energy projects, a higher rate of depreciation may be admissible, especially in the initial years, as an incentive. This gives relief to the investor in the form of income tax rebates. In some countries like US, Canada and India, wind energy investors are allowed to deduct a fraction of the capital from their tax burden. In some cases, direct cash grants are also being extended to wind energy projects. Denmark and Germany are two examples.

The major way of extending production incentive is the Production Tax Credit (PTC), which is being successfully implemented by the US Government. Under this, the electricity produced during the first ten years of operation of a wind energy project, which is sold in wholesale to the utility or other electricity suppliers, is entitled for a production tax credit. Rate of PTC is revised in tune with the inflation. Currently this is 1.8 cents/kWh. Initially this credit was meant only upto December 2003, which is being extended upto December 2006 [5]. Inorder to encourage private investors to build wind farms and produce as much energy as possible from this sustainable source, some countries like Denmark and Germany give direct cash payment to the investors on the basis of kWh produced.

Policy initiatives by the local and national governments are another important factor. In several countries, wind turbines and its accessories are exempted from customs duty. In some cases, the property tax on the land occupied by wind farms is completely or partially waived. Ensuring the finance for the project at liberal interest rate is another way of expressing support for wind energy. An example is the Alternate Energy Revolving Loan programme run by the Iowa Energy Centre (IEC). Under this initiative, half of the project cost is given by IEC in the form of interest free loan upto 20 years. The rest is arranged from financial institutions at a negotiable rate [4].

Several developed countries have formulated environmental regulations, which will ultimately come in favor of non-polluting sources like wind. For instance, emission taxes enforced in several countries demands the polluting industries to pay tax, in proportion to the quantity of CO_2 and SO_X they emit to the atmosphere. This will increase the economic competitiveness of renewable technologies. The nature and level of incentives offered to wind energy varies from region to region [1, 6].

7.2 The 'present worth' approach

The basic motive of any investment in a capitalistic society is to earn more money. Value of money varies with time. For example, if we have \$100 today and deposit it in a bank offering an interest rate of say 10 per cent, after one year our deposit

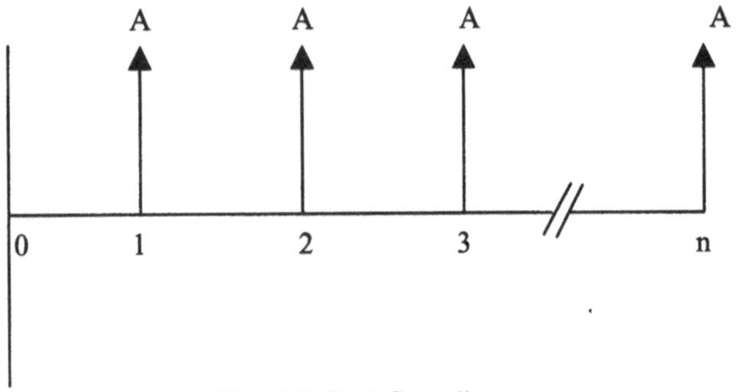

Fig. 7.5. Cash flow diagram

will grow to \$110, making us richer by \$10. Thus, today's \$100 is worth \$110 after one year. In other words, \$ 110 available after one year is worth only for \$100 today.

Wind energy projects are investments lasting for 20 to 30 years. There are cash inflows and outflows during all these years in the form of benefits and costs related to the project. Thus, for getting the real picture of the project economics, costs and benefits over the entire life span of the project has to be considered.

For assessing the project economics, cash-flows in different years of the project are brought to a common reference time. In other words, we have to assess the total worth of the project, by accumulating all cash flows to a common reference period. This reference time may conveniently be taken as the project's starting period.

Future value of an investment C made today is given by;

$$A_1 = C\,(1+i)\ ,\ A_2 = C\,(1+i)^2,\ A_3 = C\,(1+i)^3, \text{L}\ ,\ A_n = C\,(1+i)^n \tag{7.1}$$

where A_1, A_2, A_3,--- , and A_n indicate the value in the 1st, 2nd, 3rd, ---, and nth year respectively. Here i is the interest rate or as more commonly termed, the discounting rate. In other words, the present value of a receipt after n years (A_n) is given by

$$PV(A) = \frac{A_n}{(1+i)^n} \tag{7.2}$$

Now, let us consider a uniform annual cash flow for n years as shown in Fig. 7.5. Present values of payment A in different years, discounted to the initial year (year 0) is

$$PV(A)_1 = \frac{A}{(1+i)},\ PV(A)_2 = \frac{A}{(1+i)^2},\ PV(A)_3 = \frac{A}{(1+i)^3},\ \text{L}\ , \tag{7.3}$$

$$\text{and}\ \ PV(A)_n = \frac{A}{(1+i)^n}$$

Thus, the accumulated present value of all the payments put together is

$$PV(A)_{1-n} = A\left(\frac{1}{(1+i)} + \frac{1}{(1+i)^2} + \frac{1}{(1+i)^3} + \text{L} + \frac{1}{(1+i)^n}\right) \tag{7.4}$$

Eq. (7.4) can be brought to a standard geometric series by taking 1/(1+i) common. Thus,

$$PV(A)_{1-n} = \frac{A}{(1+i)}\left(1 + \frac{1}{(1+i)} + \frac{1}{(1+i)^2} + \text{L} + \frac{1}{(1+i)^{n-1}}\right) \tag{7.5}$$

which can be further reduced to

$$PV(A)_{1-n} = \frac{A}{(1+i)} \frac{1-\left(\frac{1}{(1+i)}\right)^n}{1-\left(\frac{1}{(1+i)}\right)} \tag{7.6}$$

The above expression can be simplified as

$$PV(A)_{1-n} = A\left[\frac{(1+i)^n - 1}{i\,(1+i)^n}\right] \tag{7.7}$$

Example

A wind turbine generates 1576800 kWh in an year. The generated electricity is sold to the utility at a rate of 5 cents/kWh. The discount rate is 5 per cent. Calculate the present worth of electricity generated by the turbine throughout its life period of 20 years.
Yearly revenue from the project is

$$1576800 \times 0.05 = \$78840$$

The cash flow during the 20 years is shown in Fig. 7.6.

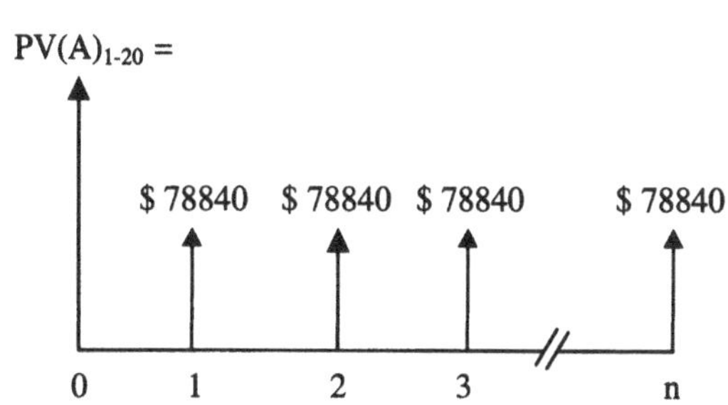

Fig. 7.6. Cash flow from electricity sales

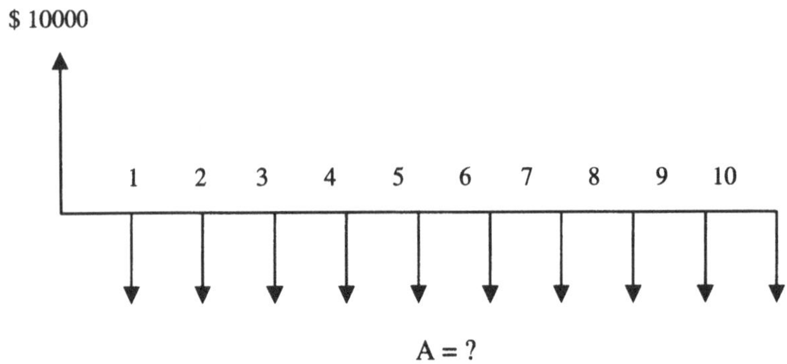

Fig. 7.7. Cash flow of loan repayment

Accumulated present value of electricity generated is given by

$$PV(A)_{1-20} = 78840 \times \left[\frac{(1+0.05)^{20} - 1}{0.05\,(1+0.05)^{20}} \right] = \$\,982521$$

Example

An amount of $ 10,000 was borrowed at a discount rate of 7 per cent and invested in a project. The loan has to be settled in 10 years through uniform annual repayment. Calculate the amount to be paid annually.

Here we have to find out the annual repayment of the loan. The cash flow is shown in Fig. 7.7.

From Eq. (7.7), the annual payment is given by

$$A = PV(A)_{1-n} \left[\frac{i(1+i)^n}{(1+i)^n - 1} \right] \tag{7.8}$$

That is

$$A = 10000 \left[\frac{0.07\,(1+0.07)^{10}}{(1+0.07)^{10} - 1} \right] = \$1424$$

In the examples illustrated above, we have considered the nominal interest or discount rate for computing the present value of the investment. Another major factor to be considered is the inflation. Physically, inflation in an economy can be considered as the increase in prices of goods and services over a period of time. Thus inflation decreases the value of money. The Consumer Price Index (CPI) commonly gauges the rate of inflation in an economy.

There are two different ways for incorporating the effect of inflation in our calculations. We can either treat the inflation and interest separately or combine the inflation with the interest and use the adjusted discount rate.

The discount rate corrected for inflation is termed as the real rate of discount (interest). We will use the real rate of discount in our calculations. The real discount rate (I) can roughly be taken as the difference between the nominal interest rate and inflation. More precisely, I is given by

$$1+I = \frac{1+i}{1+r} \tag{7.9}$$

where r is the rate of inflation.

The price of some commodities may increase at a rate higher than that of inflation. Electricity is a good example. This increase, which is over and above inflation, may be caused by many factors like scarcity of fuels, international market pressure or political and policy changes. For example, during the years of gulf war, oil prices increased rapidly, which inturn reflected as the rise in electricity tariff. The increase in the cost of a commodity, in comparison with the general inflation is termed as the escalation (e).

When the escalation rate e is combined with the inflation r, it is termed as the apparent escalation rate e_a, which is given by.

$$e_a = \{(1+e)\,(1+r)\} - 1 \tag{7.10}$$

The real rate of discount, adjusted for both the inflation and escalation is then given by

$$I = \frac{1+i}{1+e_a} - 1 \tag{7.11}$$

Example

Calculate the annual repayment in the above example, if the rate of inflation is 3 per cent.

The real discount rate, corrected for inflation is

$$I = \left[\frac{1+0.07}{1+0.03}\right] - 1 = 0.039$$

Therefore the annual payment is

$$A = 10000\left[\frac{0.039(1+0.039)^{10}}{(1+0.039)^{10} - 1}\right] = \$1227$$

Hence, when we include the effect of inflation, the annual payment is reduced to \$1227.

7.3 Cost of wind energy

There are three different ways in which the cost of a wind energy system is commonly expressed. They are (1) cost per rated power of the turbine (2) cost per unit rotor size and (3) cost per kWh of electricity generated.

The easiest way to express the cost is in terms of the rated capacity. If C_T is the cost of the turbine with a rated power P_R, then the cost per kW (C_{PR}) is given by

$$C_{PR} = \frac{C_T}{P_R} \tag{7.12}$$

Although this method is useful in comparing the market rates of systems with similar rated capacity, quite often, it may be misleading. First of all, one should ensure the components and accessories that are included in C_T as different manufactures quote the rates in different ways – some may include the turbine and all the accessories in their price, where as others quote only the turbine cost.

Another problem is that the rated power of a machine is a function of its rated wind speed. For example, we may have two systems with the same rotor area but different rated speeds; say 12 m/s and 15 m/s. As the ratio of their rated speeds is 15/12, the rated power of the second system is almost twice as that of the first one. The second system should have a bigger generator. As the generator and allied components constitute only a portion of the system's cost, there will not be much difference in C_T in the two cases. Hence, in terms of C_{PR}, we may feel the second system is cheaper. However, in a wind regime of low strength, the first system may produce more power as it can interact more efficiently with the wind spectra. Thus the first system having higher cost/kW may prove to be more economical in reality.

A better way to express the cost of the system is in terms of unit rotor size. If A_T is the rotor area, then the cost per unit size of the turbine C_A is expressed as

$$C_A = \frac{C_T}{A_T} \tag{7.13}$$

Here we consider only the rotor size in expressing the cost. Two turbines with the same rotor area may have different rated wind speeds as in the above case. In a wind regime with stronger winds, the system with higher P_R will deliver more power and hence will have economic advantage over the other.

We are often more concerned about the cost of generating a kWh of electricity from our system. Hence the cost/kWh is a better economic indicator. The characteristic wind regime is a critical factor in deciding the cost of wind-generated electricity. If C_A is the cost of operation and E_I is the kWh generated, both expressed in annual basis, then the unit cost of wind generated-electricity is

$$C_E = \frac{C_A}{E_I} = \frac{C_A}{8760\ C_F\ P_R} \tag{7.14}$$

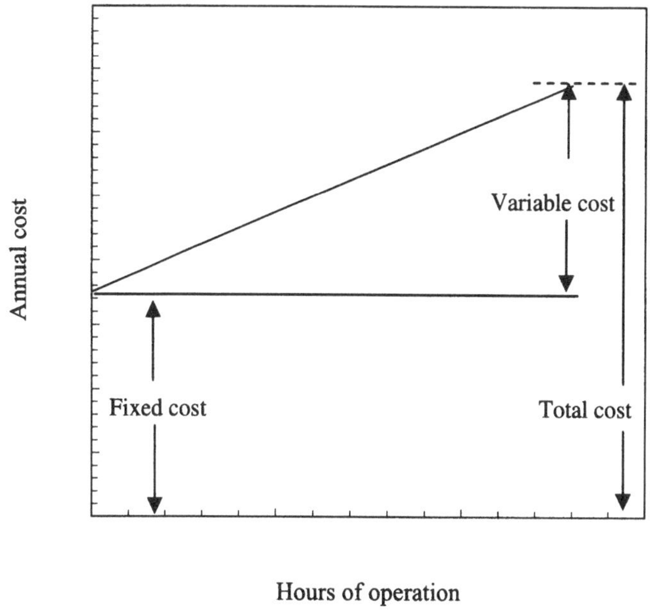

Fig. 7.8. Fixed and variable costs

where C_F is the capacity factor, which has been discussed in Chapter 5.

Annual cost of operation of a project would essentially have two components-fixed costs and variable costs. Fixed costs refer to the costs that are to be incurred by virtue of mere existence of the project. That is, fixed costs are to be met, irrespective of whether the project is functional and how much power is being generated. On the other hand, the variable costs vary in proportion to the quantity of project output. As illustrated in Fig. 7.8, the total annual cost of operation (C_A) is the sum of annual fixed costs (F_C) and variable costs (V_C). Thus,

$$C_A = F_C + V_C \tag{7.15}$$

In case of a wind energy projects, the fixed component of annual cost is contributed by the initial investment. Similarly, the variable costs consists the expenditure on the operation and maintenance of the system.

7.3.1 Initial investment

While estimating the project capital, apart from the cost of the wind turbine, costs associated with other essential components should also be considered. This may include the costs of land, extra tower (if required), controls, power conditioning unit, civil works, electrical infrastructure for grid integration and the installation charges. If the turbine and its accessories are not indigenously available, then the cost of procuring these items through import is also to be considered.

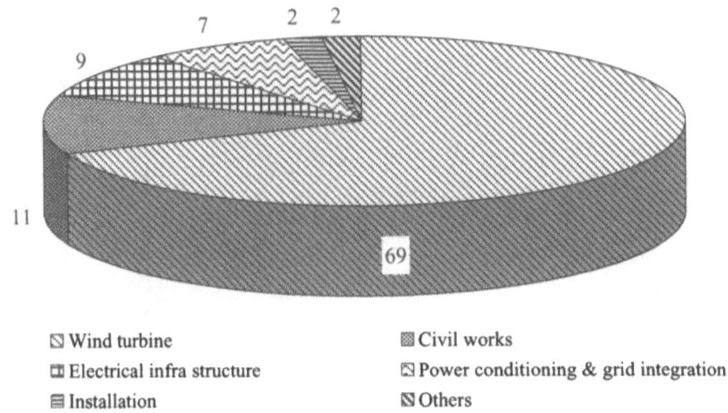

Fig. 7.9. Capital investment for a wind energy project

Table 7.1. Cost of wind turbines based on the rated power

Size (kW)	Cost, $/kW
Up to 10	2400-3000
100	1250-2000
250 and above	700-1000

Approximate market prices of wind turbines of different sizes are given in Table 7.1.

As we see, the cost/kW decreases with the increase in the system size. For machines sized above 250 kW, the turbine cost may roughly be taken as $750 per kW. Break-up of the capital cost required for a typical wind power project of 5 MW capacity is shown in Fig. 7.9. The major part of the investment-around 69 per cent- is for the turbine itself. Civil works, electrical infrastructure and power conditioning required 11, 9 and 7 per cents respectively of the total initial investment. Installation and other miscellaneous charges account for 4 per cent. It may be noted that the land costs are not included in the capital investment shown above.

The project requires a considerable area of land for installing the turbines. However, the area actually occupied by the turbines is only a small fraction of the land used for the project. Hence the free land available may be used for other purposes like agriculture, provided it does not adversely affect the flow of wind. The land required might either be purchased or taken in lease for a specified period. For quick calculations, we may take all these expenditures, clubbed together, as 30 per cent of the turbine cost.

7.3.2 Operation and maintenance costs

In conventional power plants, the major portion of the operation and maintenance cost is spent for fuel and lubrication charges. For wind energy generation, the fuel - wind- is free. However, the turbine requires periodic attention and proper maintenance for trouble free operation. Some components of the wind machine are

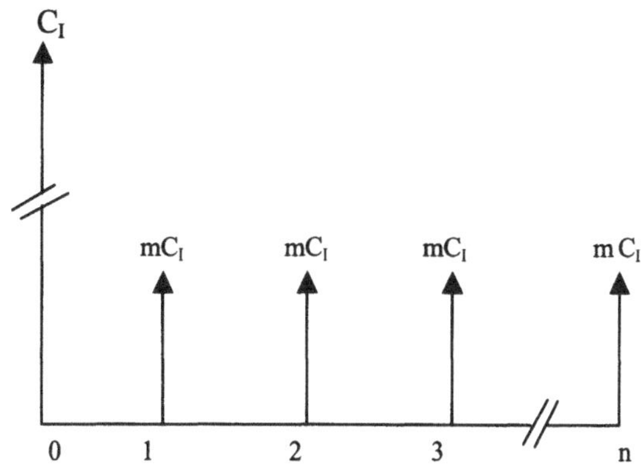

Fig. 7.10. Costs of the wind energy project

prone to wear and tear due to continuous operation; examples are gearboxes and transmission elements. Extreme aerodynamic loading may cause fatigue to the rotor blades. The frequency, and thus the cost, required for the maintenance is proportional to the hours for which the system is put to work.

The climatic factors may also affect the degree of maintenance required for the system. If the wind at the site is more turbulent, the turbine may demand more attention and care. Presence of corrosive substances in the atmosphere may reduce the life of some components.

Insurance and taxes are the other expenses being incurred annually. The project has many costly and sensitive components, which are to be insured against unforeseen accidents and calamities. The insurance agencies have a wide variety of plans to suit the needs of their customers. Here also, local enquiry will help us to choose the right one for the project. Wind energy projects are partially or fully exempted from taxes in many countries. The level of exemption depends on the local tax laws. If the land required for the project is rented, it can be included in the operation and maintenance cost. Salaries of the project staff are another expenditure coming under the operation and maintenance head. Considering all these factors, it is a usual practice to consider the operation and maintenance charges as a fraction of the capital cost of the system. It is logical to assign 1.5 to 2 per cent of the system cost for yearly repair and maintenance.

7.3.3 Present value of annual costs

Annual costs involved in a wind energy project over its life span of n years are shown in Fig.7.10. Let C_I be the initial investment of the project and C_{OM} be the operation and maintenance cost. Expressing the C_{OM} as a percentage m of C_I,

$$C_{OM} = m\,C_I \tag{7.16}$$

Now, discounting the operation and maintenance costs for n years to the initial year,

$$PV(C_{OM})_{1-n} = mC_I \left[\frac{(1+I)^n - 1}{I\,(1+I)^n}\right] \tag{7.17}$$

Including the initial investment C_I, the accumulated net present value of all the costs is represented as

$$NPV(C_A)_{1-n} = C_I \left\{1 + m \left[\frac{(1+I)^n - 1}{I\,(1+I)^n}\right]\right\} \tag{7.18}$$

Hence, the yearly cost of operation of the project is

$$NPV(C_A) = \frac{NPV(C_A)_{1-n}}{n} = \frac{C_I}{n}\left\{1 + m \left[\frac{(1+I)^n - 1}{I\,(1+I)^n}\right]\right\} \tag{7.19}$$

If P_R is the rated power of the turbine and C_F is the capacity factor, the energy generated by the turbine in an year is

$$E_I = 8760\ P_R\ C_F \tag{7.20}$$

Thus, the cost of kWh wind-generated electricity is given by

$$c = \frac{NPV(C_A)}{E_I} = \frac{C_I}{8760\ n}\left(\frac{1}{P_R\ C_F}\right)\left\{1 + m \left[\frac{(1+I)^n - 1}{I\,(1+I)^n}\right]\right\} \tag{7.21}$$

Example

Cost of a 600 kW wind turbine is \$ 550000. Other initial costs including that for installation and grid integration are 30 per cent of the turbine cost. Useful life of the system is 20 years. Annual operation and maintenance costs plus the land rent come to 3.5 per cent of the turbine cost. Calculate the cost of generating electricity from the turbine when it is installed at a site having a capacity factor of 0.25. The real rate of interest may be taken as 5 per cent.

Here, the installation cost of the turbine is

$$550000 \times \frac{30}{100} = \$165000$$

So the total initial investment for the project is 550000+165000 = \$ 715000. Hence, the cost of one kWh of electricity is

$$c = \frac{715000}{8760 \times 20}\left(\frac{1}{600 \times 0.25}\right)\left\{1 + 0.035\left[\frac{(1+0.05)^{20} - 1}{0.05\,(1+0.05)^{20}}\right]\right\} = \$0.04/\text{kWh}$$

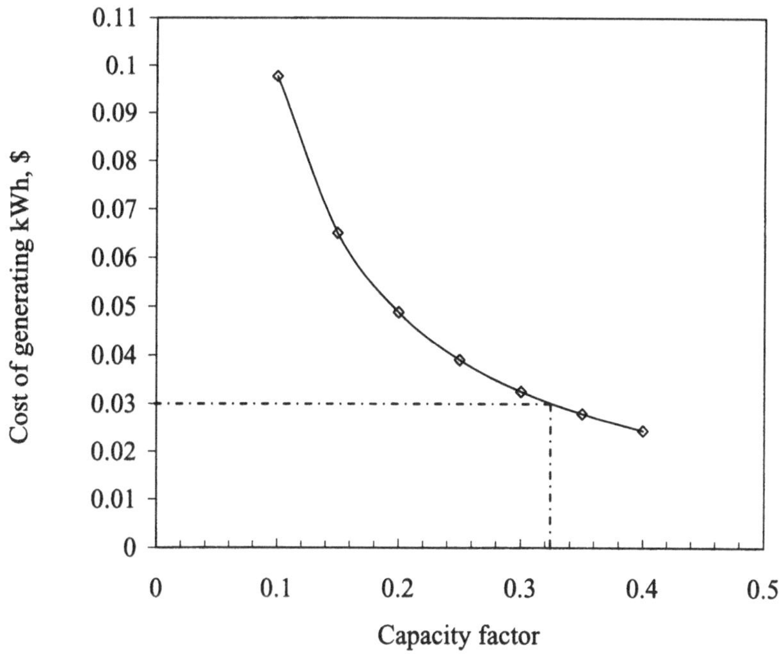

Fig. 7.11. Effect of capacity factor on energy cost

Example

Illustrate the effect of capacity factor on the cost/kWh in the above example. If a utility company buys the generated electricity at a rate of $ 0.03 /kWh, find out the break-even capacity factor.

Effect of capacity factor on the unit cost of electricity is shown in Fig. 7.11. The break-even capacity factor is the C_F for which the generating costs are equal to the selling cost. Hence we have

$$0.03 = \frac{715000}{8760 \times 20}\left(\frac{1}{600 \times C_F}\right)\left\{1 + 0.035\left[\frac{(1+0.05)^{20}-1}{0.05\,(1+0.05)^{20}}\right]\right\}$$

Solving for C_F, we get the break even capacity factor as 0.33.

7.4 Benefits of wind energy

What is the real value of electricity generated by a wind turbine? Though it sounds simple, the question is a bit tricky one. We can look at the value of electricity generated in different perspectives which makes the assessment a complex process.

If the benefits are evaluated from a nation's point of view, apart from the mere value of energy produced, the benefits of generating power through a clean and environment-friendly technology should be quantified and added to the price of wind electricity. Social costs involved in other technologies, which are hidden, also add to the benefits of wind energy.

If we evaluate the benefits from the viewpoint of a utility company, the value of wind electricity depends greatly on the time of the day at which it is available. For a private investor, who sells the generated energy to the local grid, the selling price decides his benefits. The turbine owner may consume the energy generated from wind-partially or fully-for meeting his own energy demand. As other wise he has to buy electricity from the grid at the retail price, the benefits of wind energy may be assessed on the basis of energy tariff prevailing in the region. Here, the benefits vary significantly from place to place. For example, electricity tariffs are higher in Germany than in UK and hence wind electric generation may be more profitable in Germany.

Benefits of wind energy can also be evaluated on the basis of the cost of generating energy through other technologies, which are being replaced by the wind turbine. For example, if the wind turbine replaces a diesel generator, the cost of generating energy using the diesel may be considered as the value of wind-generated electricity, in a kWh basis.

For our calculations, let us consider the retail energy price as the value of wind-generated electricity. If the project delivers a benefit of B_A annually through electricity sales, then the accumulated present value of all benefits over the life of the project is

$$NPV(B_A)_{1-n} = B_A \left[\frac{(1+I)^n - 1}{I(1+I)^n} \right] \tag{7.22}$$

Example

Calculate the net present worth of electricity sales from the wind turbine described in the above example. Electricity price is \$0.045 per kWh. Take the interest rate as 7 per cent, inflation as 3 per cent and escalation as 2 per cent.
The apparent rate of escalation, corrected for inflation is

$$e_a = \{(1+0.02)\,(1+0.03)\} - 1 = 0.05$$

Therefore the real rate of discount, adjusted for inflation and escalation is

$$I = \frac{1+0.07}{1+0.05} - 1 = 0.02$$

With the rated capacity 600 kW and the capacity factor 0.25, annual energy production is

$$E_I = 8760\, P_R\, CF = 8760 \times 600 \times 0.25 = 1314000 \text{ kWh}$$

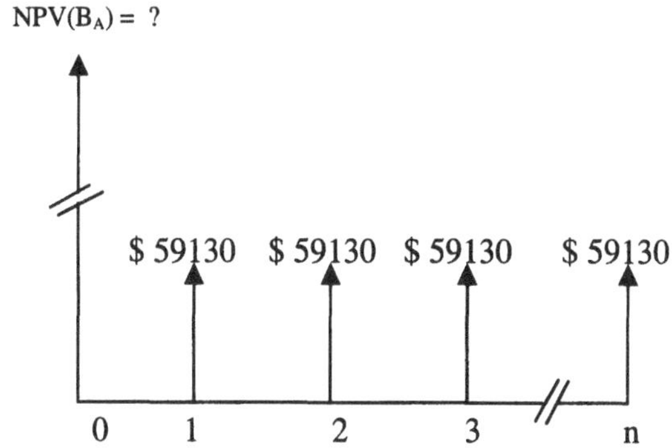

Fig. 7.12. Benefits of wind energy

Annual revenue from the sale of electricity at a rate of $ 0.045/kWh is

$$1314000 \times 0.045 = \$59130$$

The cash flow is shown in Fig. 7.12. Discounting the benefits of all 20 years to the project beginning at real discount rate and adding together, we have

$$B_A = 59130\left[\frac{(1+0.02)^{20}-1}{0.02(1+0.02)^{20}}\right] = \$\ 966860$$

7.5 Yardsticks of economic merit

The cash flow diagram of a wind energy project is shown in Fig. 7.13. As we can observe, apart from the initial investment, there are annual cash inflows and outflows associated with the project throughout its life span. Hence, looking at the project in totality, we may be further interested to know:

1. What is the present worth of the entire project?
2. Whether the benefits from the project are in proportion with the costs involved?
3. How many years will it take to get our investment back from the project?
4. What is real return of the project or the maximum rate of interest at which we can arrange capital for the project?

The indices used to address these issues are (1) Net present value (2) Benefit cost ratio (3) Pay back period and (4) Internal rate of return.

7.5.1 Net present value

The net present value (NPV) is the net value of all benefits (cash inflows) and costs (cash outflows) of the project, discounted back to the beginning of the

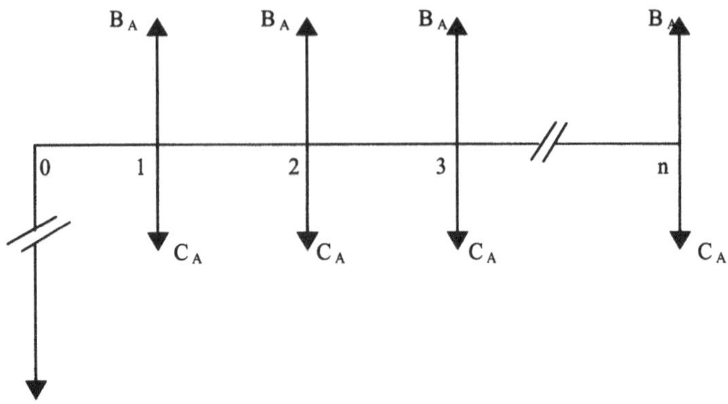

Fig. 7.13. Cash flow diagram of a wind energy project

investment. The benefits will essentially include the income from sale of electricity generated. The capital investment plus the accumulated value of annual operation and maintenance costs constitute the payments. Thus the net present value is given by

$$NPV = NPV(B_A)_{1-n} - \{\ C_I + NPV(C_A)_{1-n}\ \} \tag{7.23}$$

Substituting for $NPV(B_A)_{1-n}$ and $NPV(C_A)_{1-n}$, we have

$$NPV = B_A \left[\frac{(1+I)^n - 1}{I\,(1+I)^n}\right] - \left\{ C_I \left[1 + m \left(\frac{(1+I)^n - 1}{I\,(1+I)^n}\right)\right]\right\} \tag{7.24}$$

If the NPV is greater than 0, the project is economically acceptable as it will bring profit to the investor. While comparing investment options which are mutually exclusive, the project with higher NPV should be selected.

7.5.2 Benefit cost ratio

While comparing two projects demanding different levels of initial investments, judging their merit merely on the basis of NPV may be misleading. The project involving higher investment may tend to show an impressive NPV than the one requiring lower capital. Under such situations, Benefit Cost Ratio (BCR) is a better tool to judge the economic viability.

Benefit cost ratio (BCR) is the ratio of the accumulated present value of all the benefits to the accumulated present value of all costs, including the initial investment.

$$BCR = \frac{NPV(B_A)_{1-n}}{C_I + NPV(C_A)_{1-n}} \tag{7.25}$$

Thus we have

$$BCR = \frac{B_A \left[\dfrac{(1+I)^n - 1}{I(1+I)^n} \right]}{C_I \left[1 + m \left(\dfrac{(1+I)^n - 1}{I\,(1+I)^n} \right) \right]} \tag{7.26}$$

A project is acceptable if BCR is greater than 1.

7.5.3 Pay back period

Pay back period (PBP) is the year in which the net present value of all costs equals with the net present value of all benefits. Hence PBP indicates the minimum period over which the investment for the project is recovered. At PBP,

$$NPV\ (B_A)_{1-n} = C_I + NPV\ (C_A)_{1-n} \tag{7.27}$$

That is:

$$B_A \left[\frac{(1+I)^n - 1}{I(1+I)^n} \right] = C_I \left[1 + m \left(\frac{(1+I)^n - 1}{I\,(1+I)^n} \right) \right] \tag{7.28}$$

The pay back period is computed by solving the above equation for n. The above equation can be rewritten as

$$\frac{C_I}{B_A - mC_I} = \left[\frac{(1+I)^n - 1}{I\,(1+I)^n} \right] \tag{7.29}$$

which can be rearranged and reduced as

$$(1+I)^n = \left(1 - \frac{I\,C_I}{B_A - mC_I} \right)^{-1} \tag{7.30}$$

Taking logarithm on both sides,

$$n\,ln(1+I) = -\,ln\left(1 - \frac{I\,C_I}{B_A - mC_I} \right) \tag{7.31}$$

Thus,

$$n = -\frac{ln\left(1 - \dfrac{I\,C_I}{B_A - mC_I} \right)}{ln(1+I)} \tag{7.32}$$

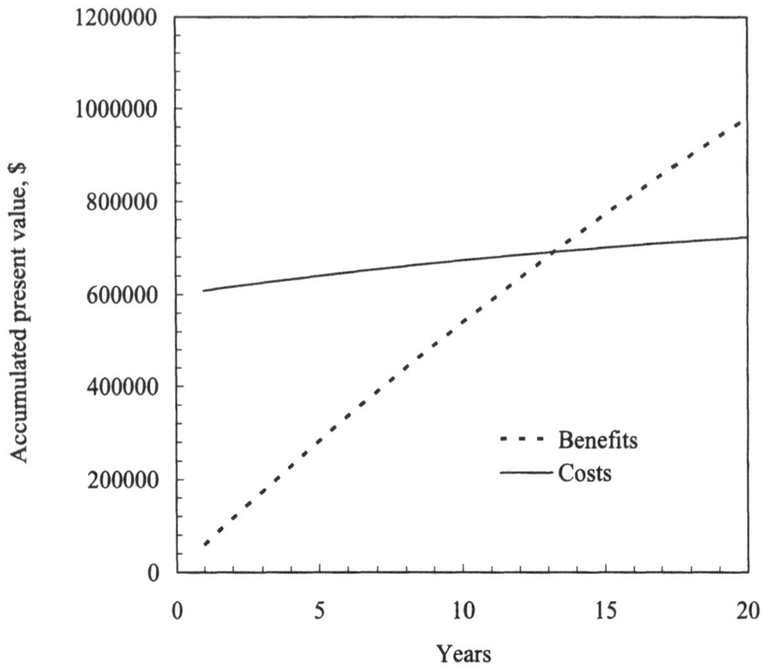

Fig.7.14. Accumulated present value of costs and benefits

The accumulated net present value of costs and benefits for a typical wind energy project is shown in Fig. 7.14. The cost curve and the benefit curve are meeting at a point corresponding to the pay back period. Here the project recovers its capital between its 13th and 14th year of operation. For obvious reasons, project with lower pay back period are preferred.

7.5.4 Internal rate of return

One of the important criteria in deciding the economic merit of a project is its internal rate of return. The internal rate of return (IRR) is defined as the discount rate at which the accumulated present value of all the costs becomes equal to that of the benefits. In other words, with IRR as the discount rate, the net present value of a project is zero.

IRR is the maximum rate of interest (in real terms) that the investment can earn. In physical sense, it is the interest rate upto which we can afford to arrange the capital for the project. As the net present value of the project discounted at IRR is zero, the PBP at IRR is the project's life period. With IRR as the discounting rate,

$$NPV(B_A)_{1-n} = C_I + NPV(C_A)_{1-n} \tag{7.33}$$

That is

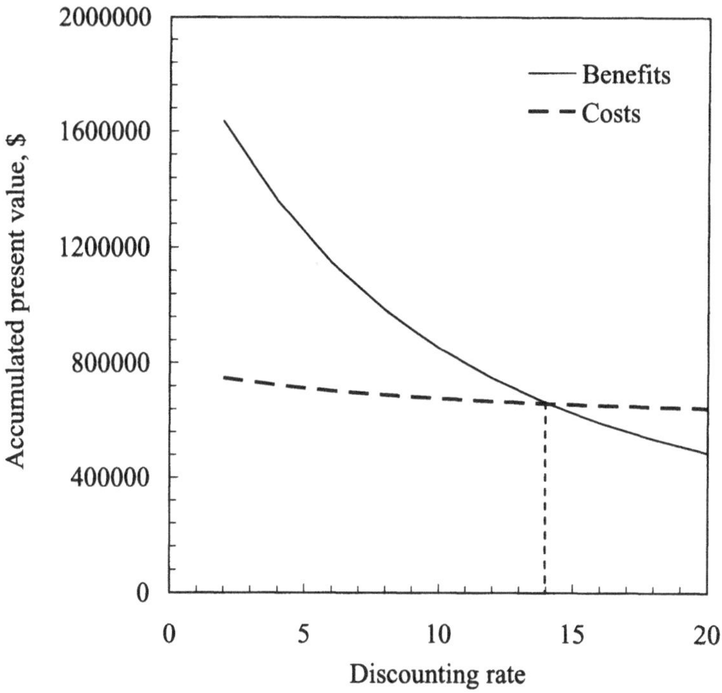

Fig.7.15. Accumulated present value of costs and benefits discounted at different rates

$$B_A \left[\frac{(1+IRR)^n - 1}{IRR(1+IRR)^n} \right] = C_I \left[1 + m \left(\frac{(1+IRR)^n - 1}{IRR\,(1+IRR)^n} \right) \right] \tag{7.34}$$

From the above expression, IRR can be solved by trial and error method or more precisely, using numerical techniques like the Newton-Raphson method.

The accumulated net present values of costs and benefits of a typical project are shown in Fig. 7.15. The intersecting point of the cost and benefit curves corresponds to the IRR of the project. In the present case it is 14 per cent, which indicates that the project is acceptable upto a real interest rate of 14 per cent.

Example

A wind energy project of 2.4 MW installed capacity requires a capital investment of $ 2200000. The site's capacity factor is 0.35. Annual operation and maintenance costs are 2.0 per cent of the initial investment. With 5 per cent real rate of discount, calculate (1) Net Present Value (2) Benefit Cost Ratio (3) Pay Back Period and (4) Internal Rate of Return. The useful project life is 25 years and the local electricity price is $0.05/kWh.
Annual electricity production expected from the project is

$$E_I = 8760 \times 0.35 \times 2400 = 7358400 \text{ kWh}$$

At the rate of \$ 0.05/kWh, the annual return from the electricity sales is

$$\mathrm{B_A} = 7358400 \times 0.05 = \$367920$$

Net present value of the benefits

$$NPV\ (B_A)_{1-25} = 367920 \times \left[\frac{(1+0.05)^{25}-1}{0.05 \times (1+0.05)^{25}}\right] = \$5185444$$

Annual operation and maintenance cost

$$\mathrm{C_A} = 2200000 \times 0.02 = \$44000$$

Net present value annual operation and maintenance cost

$$NPV(C_A)_{1-25} = 44000 \times \left[\frac{(1+0.05)^{25}-1}{0.05\ (1+0.05)^{25}}\right] = \$620134$$

Therefore the Net Present Value of the project is

$$\mathrm{NPV} = 5185444 - (620134 + 2200000) = \$2365311$$

Similarly, the benefit cost ratio is

$$BCR = \frac{5185444}{620134 + 2200000} = 1.84$$

Payback Period of the investment is

$$n = -\frac{ln\left(1 - \dfrac{0.05 \times 2200000}{367920 - 0.02 \times 2200000}\right)}{ln(1+0.05)} = 8.5 \text{years}$$

For the Internal Rate of Return we have

$$367920 \left[\frac{(1+IRR)^{25}-1}{IRR\,(1+IRR)^{25}}\right] = 2200000 \times \left\{1 + 0.02 \times \left[\frac{(1+IRR)^{25}-1}{IRR\ (1+IRR)^{25}}\right]\right\}$$

Solving the above relationship following the Newton-Raphson method, IRR is found to be 13.7 per cent.

7.6 Tax deduction due to investment depreciation

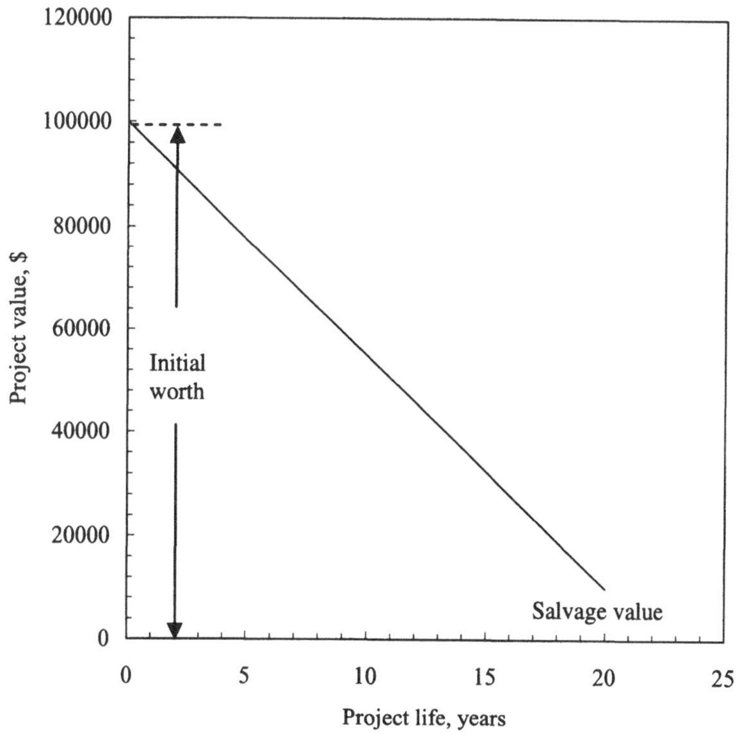

Fig. 7.16. Depreciation of a 50 kW wind turbine

Tax laws prevailing in many countries allow the investors to deduct the depreciation of their investment from the annual tax. Depreciation refers to the reduction in value of an item over its years of use. Every year, our project would depreciate at a certain rate. Value of the project at the end of its useful life period is known as the junk or salvage value.

For example, the depreciation on capital investment on a 50 kW wind turbine is shown in Fig. 7.16. The initial investment of $100000 has been reduced to its 10 per cent, at the end of 20 years of the project. This means that, every year we have to write off an amount of $ 4500 from our original asset.

There are several ways to compute the depreciation. Some extensively used methods are (1) straight line method, (2) declining balance method, and (3) sum of the years' digit method. Let us examine these methods in brief.

Straight line depreciation

In the straight-line method, we assume that the value of an item gets reduced at a uniform rate throughout its useful life period. Thus, if C_I is the capital investment for a wind turbine and S is its salvage value after its life of n years, then the annual depreciation is given by

$$D_A = \frac{C_I - S}{n} \tag{7.35}$$

For wind energy projects, the salvage value is generally taken as 10 per cent of the initial cost. It should be noted that the life considered here is the economic life period, which is generally taken as 20-30 years in case of wind turbines. Many components of the turbine may last beyond this period and with a major overhauling, the system may function for some more years.

Declining balance depreciation

In the declining balance approach, the value of an asset is written off at an accelerated rate. Thus, this method allows higher rate of depreciation in the early years of the project, than the later years. The depreciation at the end of an year is considered as a constant fraction of the book value of the project at the end of the previous year. Hence the depreciation is given by

$$D_t = p\, C_I\, (1-p)^{t-1} \tag{7.36}$$

Where D_t is the depreciation at the tth year and p is the rate of depreciation. The rate of depreciation in this method may be taken as 2/n. For example if the project life is 30 years, p is taken as 0.15. Here, the salvage value of the project is not considered.

Sum of the years' digit depreciation

We calculate the depreciation by an accelerated rate in the initial years in this method also. Thus, the depreciation at the tth year is given by

$$D_t = \frac{n-(t-1)}{n(n+1)/2}\,(C_I - S) \tag{7.37}$$

Example

The useful life of a 600 kW wind turbine costing \$ 525000 is 20 years. Compute its depreciation in the 5th, 10th and 15th years.

Let us first consider the straight-line method. Taking the salvage value as 10 per cent of the capital, the depreciation at the 5th year is given by

$$D_5 = \frac{525000 - 52500}{20} = \$23625$$

In the straight line method, depreciation for all other years is same.

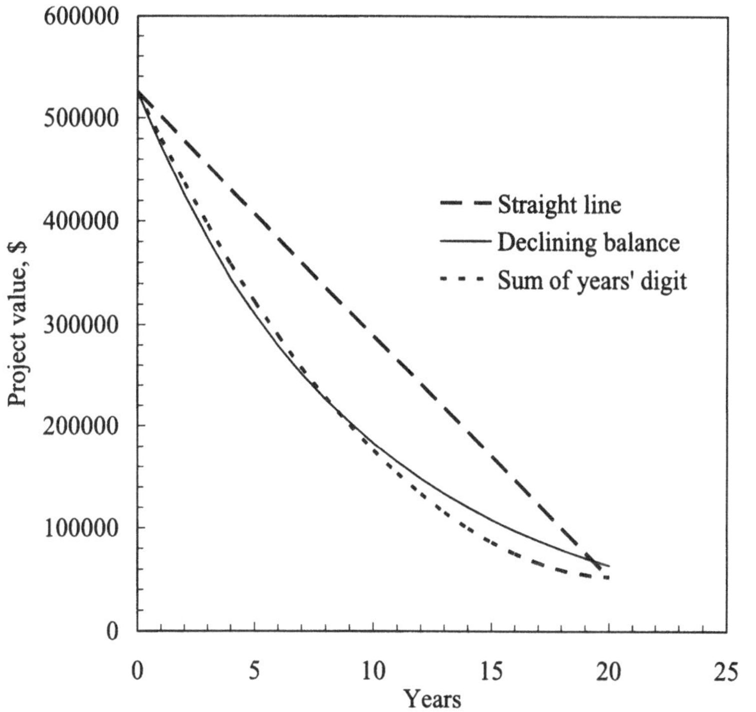

Fig. 7.17. Depreciation of the wind turbine calculated using different methods

In the declining balance method, let us take p as 2/n, which is $2/20 = 0.1$. Thus in the 5th year,

$$D_5 = 0.1 \times 525000 \times (1-0.1)^{5-1} = 34445$$

Similarly, the deprecation during 10th and 15th years are found as \$ 20340 and \$ 12010 respectively.

Under the sum of the years' digit approach, for the 5th year of the project, depreciation is given by

$$D_5 = \frac{20-(5-1)}{20(20+1)/2}(525000-52500) = 36000$$

Similarly, for the 10th and 15th years, the depreciation are \$ 24750 and \$13500 respectively.

Worth of the project at the end of each year can be computed by deducting the yearly depreciation from the value of the project at the beginning of the year. This is known as the book value of the project at that year. The book values of the turbine considered in the example, calculated using different methods of depreciation, are compared in Fig. 7.7. The sum of years' digit and declining balance methods reduce the book value of the turbine at faster rate than the straight-line method.

References

1. AWEA (1997) International wind energy incentives. American Wind Energy Association, Washington D.C.
2. AWEA (2002) The most frequently asked questions about wind energy. American Wind Energy Association, Washington D.C.
3. Fortunato B, Mummolo G, Cavallera G (1997) Economic optimization of wind power plants for isolated locations. Solar Energy 60(6):347-358
4. http://www.energy.iastate.edu
5. http://www.repp.org/repp_pubs/articles/resRpt11/preleasesubsidies.pdf
6. Lee L G (2004) Wind energy developments: Incentives in selected countries. EIA
7. Mark Z J, Gilbert M M (2001) Exploiting wind versus coal. Science 293:1438
8. Skikos G D, Machias A V (1992) FEC: A fuzzy based economic criterion for the evaluation of wind power investments. Renewable

Appendix

Wind Energy Resource Analysis (WERA) software

The Wind Energy Resource Analysis (**WERA**) programme is based on the models discussed in Chapters 3 and 5. **WERA** can be used for:
(1) Analyzing the wind energy potential at a given site.
(2) Estimating the performance of a Wind Energy Conversion System (WECS) at the site.

Present version of **WERA** has three modules, viz. **SITE, WIND TURBINE** and **WIND PUMP**. The **SITE** and **WIND TURBINE** modules have provision to perform the analysis on the basis of either Weibull or Rayleigh distribution. Wind pumping systems with positive displacement piston pump and roto-dynamic pump are considered in the **WIND PUMP** module. Procedure for installing and using the programme are discussed in the following sections.

1. Installation

The CD containing **WERA**, provided with the book, is designed to be self installing. Installation window will pop up when the CD is inserted in the drive. Proceed following the instructions to install the programme. If the auto-run does not work on your computer, install **WERA** manually by activating the setup file (**WERA\setup.exe**).

2. Wind regime analysis

For analyzing the wind energy potential at a site, activate the site module of the **WERA** programme. Depending on the type of wind data available (refer Chapter 3) we can choose either the Weibull or Rayleigh based analysis If the shape factor k and scale factor c of the site are known, select the **Weibull 'k' and 'c'** option. If we base our analysis on the raw wind data, go for **Site Wind Data.** In both the cases, number of hours for which the analysis is made has to be specified. For example, if we do our calculations on an yearly basis **Duration (h)** is 8760.

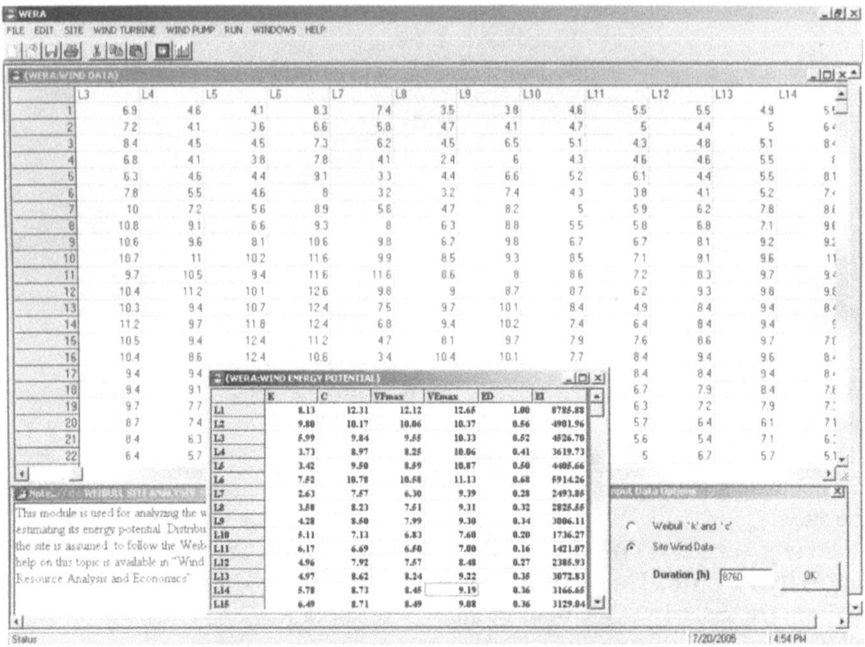

Fig. A.1. Wind resource analysis using WERA

When we click **OK** after giving the above information, **Note** window and **WERA- WIND DATA** form pop up. Values of k and c can be punched into the wind data form. Alternatively, if we have our data in any standard spread sheets like Excel, it can be copied and pasted to the WERA data form. Now, click on RUN-RESULT on the menu bar or arrow on the button bar for the results. Wind energy potential at the site, in terms of the most frequent wind velocity, velocity corresponding to the maximum energy, energy density and energy intensity, will be displayed on the **WERA: WIND ENERGY POTENTIAL** window.

If we select the **Site Wind Data** option, WERA: WIND DATA form appears. Enter or paste the site data in this form. Run the programme as discussed above to perform the wind resource analysis. WERA can analyze nine hundred and ninety nine data, from 50 locations, at a stretch. Results of the analysis can be copied and pasted to spread sheets for further manipulations. Windows of WERA under this analysis is shown in Fig. A.1.

Procedure for Rayleigh based analysis is also similar. Here, we may use either the mean wind velocity or raw wind data for our analysis.

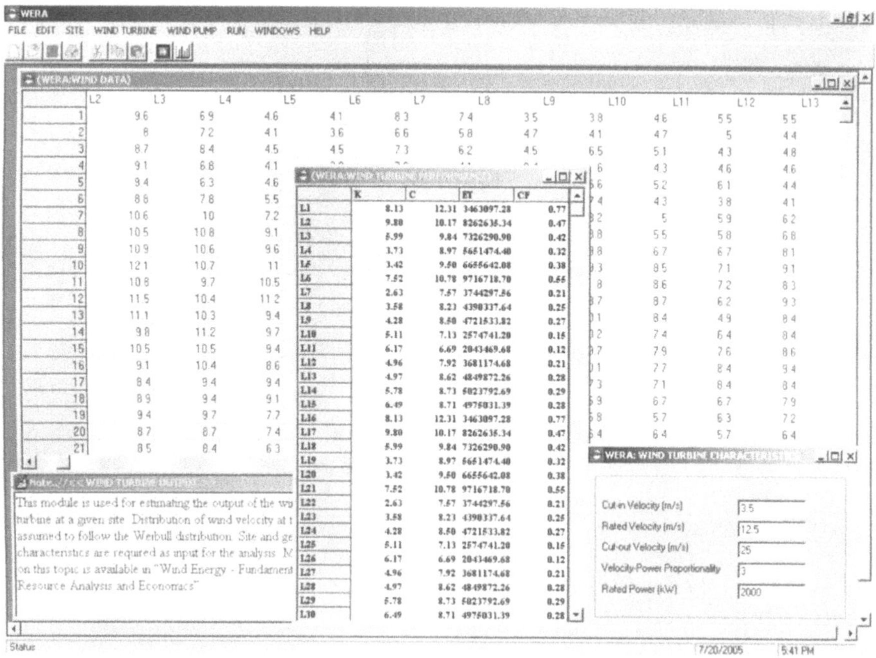

Fig. A.2. Wind turbine performance analysis using WERA

3. Wind turbine performance

WERA-WIND TURBINE module is used to assess the performance of a wind turbine in a given wind regime. This module also has the options to choose between Weibull and Rayleigh distributions for defining the site's wind profile. In the Weibull based turbine performance model also, either the site's k & c values or raw wind data can be used.

We have to specify the turbine characteristics for this analysis. Information needed are cut-in velocity, rated velocity, cut-out velocity, velocity-power proportionality and rated power of the turbine. These are defined and discussed in Chapter 5. Punch these values in the **WERA: WIND TURBINE CHARACTERISTICS** form.

Running the programme after providing the necessary input yields the **WERA: WIND TURBINE PERFORMANCE** window as shown in Fig. A.2.
Performance of the wind turbine is characterised by the total kWh generated by the system over the time period and its capacity factor. Rayleigh based analysis is also similar. Here, the input data can either be the **Average Wind Velocity** or **Site Wind Data**.

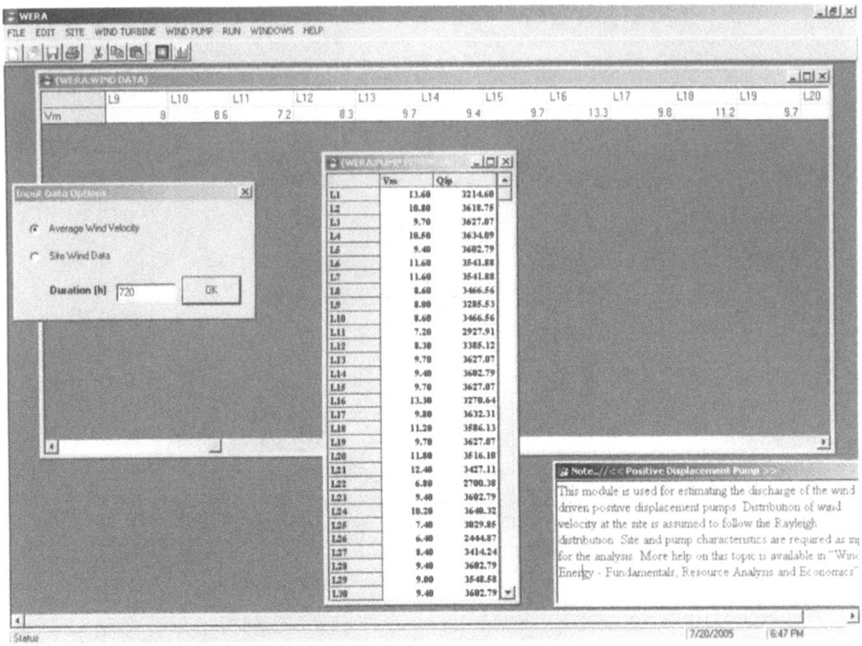

Fig. A.3. Wind pump performance analysis using WERA

4. Wind pump performance

The wind pump analysis is based on Rayleigh distribution. Under the **WIND PUMP** module, we can simulate the performances of wind driven **POSITIVE DISPLACEMENT** or **ROTODYNAMIC** pumps. As in the above cases, both the **Average Wind Velocity** and **Site Wind Data** can be used for the analysis.

Characteristics of the pumping system are to be given in the **SYSTEM PARAMETERS** form. Information required are design power coefficient of the rotor, pump efficiency, rotor diameter, constant k_0, cut-in and cut-out velocities of the wind turbine and pumping head. These parameters are discussed in Chapter 5. The output of the programme is shown in Fig. A.3.

WIND PUMP-ROTODYNAMIC module is also similar. Here, features of the system, in terms of rotor design power coefficient, combined efficiency of pump and transmission, rotor diameter, design wind velocity, gear ratio between the rotor and pump, design tip speed ratio of the rotor, pump speed corresponding to the design point, pump speed required for commencement of flow and pumping head, are to be given in the **SYSTEM PARAMETERS** form.

Index

Printing: Krips bv, Meppel
Binding: Stürtz, Würzburg

Kolibri 360 GmbH
Ferdinand-Jühlke-Straße 7
99095 Erfurt, Deutschland
produktsicherheit@kolibri360.de